T0213583

SpringerBriefs in Space Development

Series editor:
Michael K. Simpson, *Institute of Space Commerce, ISU Space Policy and Law, Boulder, CO, USA*

The SpringerBriefs in Space Development series explores the multifaceted field of space exploration and its impact on society. Under the editorship of Dr. Pelton and the auspices of the International Space University, the series features interdisciplinary contributions from space experts and rising professionals alike, offering unique perspectives on a rapidly expanding field.

The volumes in this series are compact, ranging from 50 to 125 pages (25,000–45,000 words). Instructors, students, and professionals will find herein a host of helpful reads, including snapshot reviews of a hot or emerging field; introductions to core concepts; extended research reports providing more detail than is possible in a conventional journal article; manuals describing an experimental technique or technology; and essays on new ideas and their impact on science and society.

The Briefs are concise, forward-looking studies on the space industry, broaching topics such as:

- Space technology design and optimization
- Astrodynamics, spaceflight dynamics, and astronautics
- Resource management and mission planning
- Human factors and life support systems
- Space-related law, politics, economics, and culture

Through the SpringerBriefs in Space Development, readers will learn of the amazing progress and key issues born from the international effort to explore space. The Briefs are published as part of Springer's eBook collections, and in addition are available for individual print and electronic purchase. They are characterized by fast, global electronic dissemination, straightforward publishing agreements, easy-to-use manuscript preparation and formatting guidelines, and expedited production schedules.

More information about this series at http://www.springer.com/series/10058

James Miller • Connie J. Weeks

General Relativity for Planetary Navigation

INTERNATIONAL®
SPACE UNIVERSITY

Springer

James Miller
Porter Ranch
CA, USA

Connie J. Weeks
Loyola Marymount University
Northridge
CA, USA

ISSN 2191-8171   ISSN 2191-818X   (electronic)
SpringerBriefs in Space Development
ISBN 978-3-030-77545-2   ISBN 978-3-030-77546-9   (eBook)
https://doi.org/10.1007/978-3-030-77546-9

This Springer imprint is published by the registered company Springer Nature Switzerland AG
The registered company address is: Gewerbestrasse 11, 6330 Cham, Switzerland

Contents

Chapter 1
Einstein Field Equations

The General Theory of Relativity, as it relates to navigation of spacecraft, can be separated into two parts. The first part involves derivation of a set of differential field equations that can be solved for the metric tensor. The second part involves inserting the metric tensor into the equation of geodesics to obtain equations of motion, which can be solved for formulae describing the precession of Mercury's orbit, the bending of light, radar time delay, gravitational red shift, and the time measured by clocks. In this chapter, the solution for the metric tensor is obtained from equations that provide a statement of the theory's fundamental assumptions. The assumptions are simply that the speed of light is constant, matter or energy curves space and the universe have some symmetrical properties. These assumptions are observed and cannot be proven. Two methods are used to solve for the metric tensor. The first is a computer solution that involves parameterizing the metric tensor and solving for the parameters using an orbit determination filter. The second is an analytic solution developed by Einstein by defining a covariant derivative and differentiating to obtain the Riemann tensor, Ricci tensor, and Einstein's field equations, which can be solved for the metric tensor.

1.1 Introduction

The Einstein field equations have been solved exactly for the case of spherical symmetry by Schwarzschild. This solution and Einstein's solution have spawned a number of formulae describing the precession of Mercury's orbit, the bending of light, radar time delay, gravitational red shift, and several more that relate to special relativity. The Schwarzschild solution has been transformed to a form such that the equations of motion look like Newton's equations of motion with a small relativistic perturbation. For orbit determination, these equations have been programmed into software used for navigation. One might question whether this is really necessary,

J. Miller, C. J. Weeks, *General Relativity for Planetary Navigation*, SpringerBriefs in Space Development, https://doi.org/10.1007/978-3-030-77546-9_1

since the perturbations due to general relativity are so small. The justification is that the orbit solution used for prediction of a spacecraft orbit is obtained after analysis of data residuals, the difference between the real world and the world computed by a mathematical model. Since the data is very high precision, a very small modeling error will show up as a signature in the data residuals. Without relativity modeling, a serious modeling error in another variable could be masked. A navigation analyst might initially conclude that the signature is caused by relativity or some other error source such as a clock failing to keep the right time. Eventually the signature will grow in magnitude and the alarm bells will ring indicating a problem. The earlier the problem is detected, the more likely a solution can be found before the spacecraft crashes into something. The problem of an inaccurate gravity harmonic caused an exponential rise in the Doppler signature on the Near Earth Asteroid Rendezvous (NEAR) mission which was detected early and corrected before anything catastrophic happened. For this reason, general and special relativity are programmed into the navigation operational software.

In the 1960s, general relativity was programmed into the Orbit Determination Program (ODP) at the Jet Propulsion Laboratory (JPL). At the time, those outside of navigation thought this was not needed. Since that time, many orbits have been determined using the ODP and little attention has been given to general relativity. The ODP is treated as a black box. With the advent of comet and asteroid missions a new orbit determination program was needed. This effort required implementing general relativity. Finding and understanding the equations presented a major difficulty. After consulting many sources including relativity experts at JPL, equations were programmed into the software used for the NEAR mission. We can assume the equations are correct because the spacecraft completed its mission successfully.

The derivation of the relativity equations of motion was initiated from the metric tensor which was assumed to be correct. The goal of deriving the equations from Einstein's original assumptions that the speed of light is constant and matter curves space has been difficult to achieve. The equations of motion were worked out long before Einstein's death. His theory written in books published up to that time were close to his original 1916 paper [1]. After his death, cosmologists got hold of the theory, and engineers had difficulty understanding the mathematics. The main source of confusion was the normalization of coordinates removing c, the speed of light, and G the gravitational constant from the equations. Einstein did this to make his theory look more profound and mathematical. In this chapter, the part of Einstein's theory pertaining to navigation of spacecraft in the solar system has been extracted from Einstein's original paper [1], Eddington's book [2] written in 1923, Harry Lass's book [3] on tensors written in 1950, and Sokolnikoff's book [4] written in 1951. What goes on inside the sun, earth or black holes is not relevant to navigation of spacecraft in the solar system. While Einstein's paper is difficult to understand, all the essential equations are there. Einstein's audience was other mathematicians and physicists. Eddington, who was a mathematician, explained some of the theory in a clear way that is comparatively easy to understand. His audience was much wider than Einstein's. Sokolnikoff shows how the Riemann tensor is put together and Schwarzschild's solution is obtained. Harry Lass described the properties of tensors.

1.2 Summary of General Relativity Fundamental Assumptions

The universe assumed for navigation consists of the solar system and massless stars that are infinitely far away and emit light. The center is the solar system barycenter or center of gravity. The goal is to define the equations of motion in curved space. A Euclidean coordinate system is defined far away from the sun but not as far as the stars. If we move this coordinate system so it is centered at the solar system barycenter, we can define a curved space coordinate system as a covariant mapping from Euclidean coordinates. Sometimes we are interested in curved space coordinates and other times we are interested in Euclidean coordinates. For the equations of motion, we are interested in curved space coordinates. To define the Einstein tensor, we consider volume elements that have the same size and we use Euclidean coordinates. In Euclidean coordinates the volume elements are equal and cubical. For constant density the mass of every volume element is the same. When we map to curved space, the volume elements vary in size and shape. Since a one to one mapping exists, the curved space volume elements would have to be assigned different densities to have the same mass. Einstein defines the density in Euclidean space as scalar invariant density and this is mapped to curved space to keep the total mass the same as defined in Euclidean space.

The fundamental assumptions of General Relativity are stated in equations without proof. The first assumption is that the speed of light is constant defined by c and the observed speed of light defined by the path length ds is also constant and equal to c. It is also necessary to define a measurement (Z) which is the projection of the observed acceleration of a point mass or any vector that can be observed in curved space on the trajectory of a curved line in the gravity field defined by the equation of geodesics.

$$Z = A_u \frac{dx_u}{ds}$$

This scalar measurement gives us one equation, but there are 10 independent elements in the metric tensor. To determine them we need nine more equations. For an analytic solution, we can differentiate this measurement with respect to the assumed coordinate system to obtain four more equations that can be measured. Differentiating again gives four more equations that define the curvature and can also be observed. We need one more equation to solve for the metric tensor. It is obtained by assuming the scale or curvature is proportional to mass. The assumption that the curvature of space is proportional to mass is satisfied by placing a boundary condition on the solution to the Einstein field equations or solving the Einstein tensor by equating it to the stress–energy tensor.

There are other assumptions associated with mathematics that are difficult to state in simple equations. These include symmetry, linearity, and continuity. Not only the trajectory of a particle but all the higher order derivatives must be continuous. They trace a smooth curve when drawn on graph paper and they have slopes and areas under the curve. Once the above fundamental equations are defined, the work of the scientist is complete. For a solution, we turn the problem over to mathematicians. Einstein was the essential bridge between the two camps. His main contribution besides special relativity was the Einstein tensor which is a purely mathematical result but required considerable physical insight to derive.

1.3   Geodesic Equation

The shortest distance between two points on a curved surface is called a geodesic. When an airplane flies over the North pole on its way to Europe, it is following a geodesic or great circle arc. The metric tensor (g_{uv}) defines the arc length due to curvature of space where u and v are indices in four-space corresponding to coordinates. For example, coordinates may be x, y, z, ct as u and v vary from integer values 1 to 4. The metric tensor defines a differential line element (ds). The elements of g_{uv} are functions of space and time that define g_{uv} at some point in space. The integral of the line element (ds) gives the distance between two points or the length of the curve connecting them. Consider two points A and B. A coordinate system can be used to locate the two points relative to one another. Since the reference coordinate system is arbitrary, the coordinates of the points are of little use. The only useful physical reality is the distance between the two points. The metric tensor can be integrated to determine the length of this line. Next, we consider a line between the two points that is the shortest distance. The variation of the path length with respect to the coordinates must be zero since only one path is the shortest. Thus we have

$$ds^2 = g_{uv}dx_u dx_v$$

$$2ds\delta(ds) = dx_u dx_v \delta g_{uv} + g_{uv}dx_u \delta(dx_v) + g_{uv}dx_v \delta(dx_u)$$

and

$$2ds\,\delta(ds) = dx_u dx_v \frac{\partial g_{uv}}{\partial x_\sigma}\delta x_\sigma + g_{uv}dx_u\,d(\delta x_v) + g_{uv}dx_v\,d(\delta x_u)$$

and the stationary condition is

$$\int \delta(ds) = 0$$

$$\frac{1}{2} \int \left[\frac{dx_u}{ds} \frac{dx_v}{ds} \frac{\partial g_{uv}}{\delta x_\sigma} \delta x_\sigma + g_{uv} \frac{dx_u}{ds} \frac{d}{ds} \delta x_v + g_{uv} \frac{dx_v}{ds} \frac{d}{ds} \delta x_u \right] ds = 0$$

The dummy indices on the last two terms can be changed to be in the same order as the first term. This trick is a property of summation notation and amounts to changing the order of the rows in matrices and a vector that is to be multiplied such that the vector can be factored out and the matrices summed before the multiplication. For more information on this property, consult Einstein's 20 pages on tensor algebra in his 1916 paper or Harry Lass's book on tensors. Here, we perform the operation and rely on the references by Einstein and Harry Lass to obtain

$$\frac{1}{2} \int \left[\frac{dx_u}{ds} \frac{dx_v}{ds} \frac{\partial g_{uv}}{\delta x_\sigma} \delta x_\sigma + \left(g_{u\sigma} \frac{dx_u}{ds} + g_{\sigma v} \frac{dx_v}{ds} \right) \frac{d}{ds} \delta x_\sigma \right] ds = 0 \qquad (1.1)$$

Integration by parts is defined by the following equation.

$$\int_{x_1}^{x_2} y \, dx + \int_{y_1}^{y_2} x \, dy = xy \Big|_{x_1 y_1}^{x_2 y_2} = x_2 y_2 - x_1 y_1$$

The geometrical interpretation of this equation is that the area between the x axis and curve between the limits x_1 and x_2 plus the area between the y axis and the curve is equal to the rectangular area $x_2 y_2$ minus the rectangular area $x_1 y_1$. The following property of differentials is provided by Einstein in his 1916 paper.

$$\frac{d}{ds}(\delta x_\sigma) = \delta \left(\frac{dx_\sigma}{ds} \right)$$

This equations follows directly from the definition of the second derivative.

In his derivation, Eddington omitted this equation probably because he thought it was trivial. Einstein included this equation because it is important to the understanding even if it looks strange. Einstein liked to make statements that are counter intuitive. Reading his paper for the words is satisfying even if the equations are not understood. If we let

$$y = \left(g_{u\sigma} \frac{dx_u}{ds} + g_{\sigma v} \frac{dx_v}{ds} \right) \qquad dy = d \left(g_{u\sigma} \frac{dx_u}{ds} + g_{\sigma v} \frac{dx_v}{ds} \right)$$

$$x = \frac{dx_\sigma}{ds} \qquad dx = \frac{d}{ds}(\delta x_\sigma)$$

Eq. (1.1) is then

$$\frac{1}{2} \int \left[\frac{dx_u}{ds} \frac{dx_v}{ds} \frac{\partial g_{uv}}{\delta x_\sigma} - \frac{d}{ds} \left(g_{u\sigma} \frac{dx_u}{ds} + g_{\sigma v} \frac{dx_v}{ds} \right) \right] \delta x_\sigma \, ds = 0 \qquad (1.2)$$

This equation must hold for all arbitrary displacements of δx_σ. If we make δx_σ extremely small, the difference between the xy rectangles defined above become negligible and can be discarded. We then make ds infinitely smaller than δx_σ and we are left with the term in the brackets. This term does not go to zero because we must add up the same infinity of ds intervals, a Riemann sum. The terms in the bracket must be zero. The line integral is an increasing monotone from point A to point B. Thus, every interval of the integrand must be zero, because if any interval is not zero, there can never be a negative interval to restore the total integration to the path length. Christoffel obtained the following result by defining an integral and then arguing that the integration must be zero, not by actually integrating. The secret to relativity theory is to define things that are zero and avoid doing any real mathematics. This approach makes the theory difficult to understand but is probably the only way the problem can be solved. Carrying out the differentiation indicated in Eq. (1.2) gives

$$\frac{dx_u}{ds}\frac{dx_v}{ds}\frac{\partial g_{uv}}{\delta x_\sigma} - \left(\frac{dg_{u\sigma}}{ds}\frac{dx_u}{ds} + \frac{dg_{\sigma v}}{ds}\frac{dx_v}{ds} + g_{u\sigma}\frac{d^2x_u}{ds^2} + g_{\sigma v}\frac{d^2x_v}{ds^2} \right) = 0$$

The chain rule applied to summation notation is

$$\frac{dg_{u\sigma}}{ds} = \frac{\partial g_{u\sigma}}{\partial x_v}\frac{dx_v}{ds} \qquad \frac{dg_{\sigma v}}{ds} = \frac{\partial g_{\sigma v}}{\partial x_u}\frac{dx_u}{ds}$$

and since the metric tensor is symmetric ($g_{v\sigma} = g_{\sigma v}$)

$$\frac{dx_u}{ds}\frac{dx_v}{ds}\left(\frac{\partial g_{uv}}{\delta x_\sigma} - \frac{\partial g_{u\sigma}}{\delta x_v} - \frac{\partial g_{v\sigma}}{\delta x_u} \right) - 2g_{e\sigma}\frac{d^2x_e}{ds^2} = 0$$

The next step is to multiply through by the contravariant metric tensor $g^{\alpha\sigma}$.

$$g^{\alpha\sigma}g_{e\sigma} = \delta_\epsilon^\alpha$$

In matrix notation, this is the same as multiplying the metric tensor by its inverse.

$$[g_{e\sigma}]^{-1}[g_{e\sigma}] = I$$

In Einstein's description of the contravariant fundamental tensor (the inverse of the covariant metric tensor), he describes the matrix inversion process which involves cofactors and determinants and the Kronecker delta which is the identity matrix. Eddington had a similar description. One advantage of summation notation is that the order of multiplication is arbitrary, so

$$\frac{dx_u}{ds}\frac{dx_v}{ds}g^{\sigma\alpha}\left(\frac{\partial g_{uv}}{\delta x_\sigma} - \frac{\partial g_{u\sigma}}{\delta x_v} - \frac{\partial g_{v\sigma}}{\delta x_u} \right) - 2\frac{d^2x_\alpha}{ds^2} = 0$$

The equation for a geodesic is thus

$$\frac{d^2x_\alpha}{ds^2} + \{uv, \alpha\} \frac{dx_u}{ds} \frac{dx_v}{ds} = 0 \tag{1.3}$$

and the Christoffel symbols are defined by

$$\{uv, \alpha\} = \frac{1}{2} g^{\sigma\alpha} \left(\frac{\partial g_{u\sigma}}{\delta x_v} + \frac{\partial g_{v\sigma}}{\delta x_u} - \frac{\partial g_{uv}}{\delta x_\sigma} \right) \tag{1.4}$$

1.4 Computer Solution for Metric Tensor

The metric tensor is symmetric and has 10 independent elements at each point in four-space. If we knew the location of 10 points in the real world, we could use the definition of the metric to solve for g_{uv}. We only know one, the vector from the sun to the spacecraft. We can get around this problem by assuming that both bodies in the real world have mass and the mass of the central body or sun is much greater than the mass of the spacecraft. Now there is a force at every point along the geodetic line that results in an acceleration of the spacecraft that can be observed. We can take the dot product of this force vector with the line element, which is in the direction of the velocity vector, and this gives us an observable measurement. We can use this measurement at various points along the path to solve for the metric tensor. We need at least 10 points and by assuming a coordinate system all the mathematics associated with Einstein's solution are bypassed.

Given the equation of geodesics, our objective is to determine the metric tensor in an assumed coordinate system. Once the metric tensor is known, we have equations of motion of the spacecraft that can be integrated twice to obtain the path which navigators call the trajectory or ephemeris. A direct approach is to parameterize the metric tensor as a function of the coordinates and solve for the parameters with an orbit determination filter. This approach is only practical if we have very high precision measurements.

Consider a spacecraft in orbit about the sun somewhere in the orbit of Mercury's orbit but far from Mercury. The estimated parameters would be the initial spacecraft state and the parameters that characterize the metric tensor. In flat space the parameters that characterize the gravity field such as gravity harmonic coefficients would replace the metric tensor. The measurements could be Doppler and range data from the Deep Space Network. Thus we have for the metric tensor

$$g_{11} = e^\phi$$

$$\phi = A_0 + A_1 r + A_2 r^2 + A_3 r^3 + \cdots$$

$$g_{22} = -r^2$$

$$g_{33} = -r^2 \sin\theta$$

$$g_{44} = e^\lambda$$

$$\lambda = B_0 + B_1 r + B_2 r^2 + B_3 r^3 + \cdots$$

and all the other elements of the metric tensor are zero as a result of spherical symmetry. This is essentially the same metric that Schwarzschild assumed for his solution except that g_{11} and g_{44} are parameterized here as a function of r. We know from symmetry that these terms must be a function of only r. The curvature of space is static so there is no time dependence. Since we do not have a spacecraft in the desired orbit, we can use the exact Schwarzschild equations of motion to simulate the spacecraft trajectory. The computed equations of motion are obtained by substituting the Christoffel symbols computed from the parameterized metric into the equation of geodesics. We thus obtain for the computed equations of motion

$$\frac{d^2 r}{ds^2} = \Gamma^1_{11}\left(\frac{dr}{ds}\right)^2 + \Gamma^1_{22}\left(\frac{d\phi}{ds}\right)^2 + \Gamma^1_{44}\left(\frac{dct}{ds}\right)^2$$

where

$$\Gamma^1_{11} = -\frac{1}{2}\frac{1}{g_{11}}\frac{\partial g_{11}}{\partial r}$$

$$\Gamma^1_{22} = -\frac{1}{2}\frac{1}{g_{11}}\frac{\partial g_{22}}{\partial r}$$

$$\Gamma^1_{44} = -\frac{1}{2}\frac{1}{g_{11}}\frac{\partial g_{44}}{\partial r}$$

and

$$\frac{d^2 r}{ds^2} = -\frac{1}{2g_{11}}\frac{\partial g_{11}}{\partial r}\left(\frac{dr}{ds}\right)^2 - \frac{r}{g_{11}}\left(\frac{d\phi}{ds}\right)^2 + \frac{1}{2g_{11}}\frac{\partial g_{44}}{\partial r}\left(\frac{dct}{ds}\right)^2$$

The line element is given by

$$ds^2 = g_{11}dr^2 + g_{22}d\phi^2 + g_{44}c^2 dt^2$$

and, for $ds^2 = c^2 d\tau^2$, where τ is the proper time, we obtain

$$\left(\frac{dt}{d\tau}\right)^2 = \frac{1}{g_{44}} - \frac{g_{11}}{c^2 g_{44}}\left(\frac{dr}{d\tau}\right)^2 - \frac{g_{22}}{c^2 g_{44}}\left(\frac{d\phi}{d\tau}\right)^2$$

The equations of motion, after substituting the metric equation, become

$$\frac{d^2r}{d\tau^2} = \frac{1}{2}\frac{1}{g_{44}}\frac{\partial g_{44}}{\partial r}\left(\frac{dr}{d\tau}\right)^2 - \frac{r}{g_{11}}\left(\frac{d\phi}{d\tau}\right)^2 - \frac{1}{2}\frac{\partial g_{44}}{\partial r}c^2$$

$$-\frac{1}{2g_{44}}\frac{\partial g_{44}}{\partial r}\left(\frac{dr}{d\tau}\right)^2 - \frac{1}{2}\frac{\partial g_{44}}{\partial r}\frac{g_{22}}{g_{11}g_{44}}\left(\frac{d\phi}{d\tau}\right)^2 \qquad (1.5)$$

$$\frac{d^2\phi}{d\tau^2} = -\frac{2}{r}\frac{dr}{d\tau}\frac{d\phi}{d\tau}$$

where

$$g_{11} = \frac{-1}{g_{44}}$$

The first and fourth terms which are functions of radial velocity cancel when g_{11} is the negative reciprocal of g_{44}. This is a result of the assumption that there is no gravity drag and acceleration does not depend on velocity. There is no ether to slow down the planets resulting in the planets falling into the sun. The third term has c^2 in the numerator. The partial of g_{44} with respect to r must have c^2 in the denominator or the equation for r acceleration will blow up or at least become very large. The third term is the Newtonian acceleration in flat space. Einstein commented on the unusual mathematical quirk that the Newtonian acceleration comes from the g_{44} term of the metric tensor. The final form of the equations of motion are

$$\frac{d^2r}{d\tau^2} = -\frac{\mu}{r^2} - \left\{\frac{r}{g_{11}} - \frac{r^2}{2}\frac{\partial g_{44}}{\partial r}\right\}\left(\frac{d\phi}{d\tau}\right)^2$$

$$\frac{d^2\phi}{d\tau^2} = -\frac{2}{r}\frac{dr}{d\tau}\frac{d\phi}{d\tau}$$

If we borrow the exact solution for g_{44} from Schwarzschild which is

$$g_{44} = 1 - \frac{2\mu}{c^2 r}$$

we get for the radial acceleration Eq. (2.17)

$$\frac{d^2r}{d\tau^2} = -\frac{\mu}{r^2} + \left(r - \frac{3\mu}{c^2}\right)\left(\frac{d\phi}{d\tau}\right)^2 \qquad (1.6)$$

Next, we insert the parameterized metric into the modeled equation of geodesics and integrate the equations of motion for a few weeks. Additionally, we could integrate the variational equations to obtain the partial derivatives of Doppler and

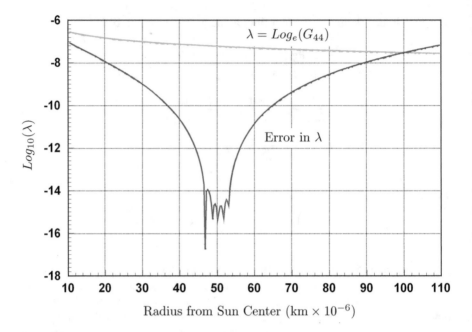

Fig. 1.1 Metric tensor estimation error

range measurements with respect to the estimated parameters, which are the metric polynomial coefficients and initial spacecraft state. To make this demonstration simple, it is assumed that we can measure $\ddot{r}$ directly and that the spacecraft trajectory is known with high precision.

The solution for the polynomial coefficients is obtained by processing several hundred data points using a weighted least square data filter. The data points are obtained from the exact Schwarzschild solution. In theory we do not need the covariant derivative, Riemann's tensor, Ricci's tensor, and Einstein's tensor to do navigation. In practice, the accuracy of the data would limit the accuracy of the parameterized metric. The result of processing an orbit of a spacecraft in Mercury's orbit is shown in Fig. 1.1. The top curve is a plot of the base 10 logarithm of the natural logarithm of g_{44} as a function of distance from the sun.

$$\lambda = -0.287656389797D - 06 + 0.112620203757D - 13r - 0.2203417321282D - 21r^2$$

$$+ 0.2154331059460D - 29r^3 - 0.8420743154620D - 38r^4$$

$$g_{44} = e^{\lambda}$$

The bottom curve is the base 10 logarithm of the difference between $\ln(g_{44})$ and the orbit determination solution for the polynomial coefficients or metric tensor. Over the range from 47×10^6 km to 57×10^6 km the fit is at the limit of computer precision. We see numerical noise at around 15 decimal places of accuracy. Over the range from 30×10^6 km to 70×10^6 km which covers the entire orbit of the spacecraft, the error is less than 1%. The precession of the orbit was 496.62 nrad per revolution about the sun which compares favorably with the Einstein formula of 479.98 nrad. The precession is about 40 arcsec per century.

1.5   Covariant Derivative of a Vector

Since the metric tensor is only a function of the distribution of matter in space and is independent of the method used to determine the orbit, it should be possible to eliminate the measurement from the differential equations for the metric tensor. Einstein and a host of mathematicians came up with a solution for eliminating the observations from the orbit determination solution and thus obtained differential equations that can be solved directly for the metric tensor. The metric tensor is obtained by placing the appropriate boundary conditions on these differential equations. This is the same idea as is used to solve Laplace's equation for gravity harmonics.

Consider the following product of a covariant vector with a contravariant vector.

$$A_u \frac{dx_u}{ds}$$

In matrix notation this would be a row vector times a column vector or the dot product. This dot product is a scalar function of the coordinates and represents a measurement of the spacecraft motion. The vector A is arbitrary in that there are many different measurements that may be used to determine the spacecraft motion. However, to make the problem simple, the vector A is projected onto the observed motion. This is equivalent to determining an orbit by observing the one-dimensional range or range rate between a spacecraft and a tracking station. For the above orbit determination solution the orbit is sampled at points along the trajectory. This would make the analytic solution difficult because it would be necessary to map the measurement in space and time. Another approach is to define alternate measurements at a point in space-time. We could measure the first and higher order derivatives of A. Whatever A is, it must be eliminated from the equations to obtain a solution for the metric tensor. Therefore, we make the observation mathematically simple and thus make eliminating it simple.

The vector A_u and the velocity or direction of motion are dependent on the assumed coordinate system. The vector A_u is not acceleration but a measure of acceleration and is thus non dynamic. The motion of a body is not dependent on the measurement of the motion. The dot product of the vector A_u with the velocity vector is independent of the assumed coordinate system. The projection of acceleration on velocity is the same if viewed from any vantage point. Imagine a flag pole mounted on a spacecraft and pointed in the direction of A_u. The spacecraft is turned to keep the solar panel pointed in the direction of the velocity vector. If the sun is in a certain direction, the length of the shadow of the flagpole on the solar panels will be equal to the projection of A_u on the velocity vector or the above dot product. The length of the shadow of the flag pole will be the same when viewed by any observer. The dot product is invariant with respect to the observed coordinates (ds). Therefore, the derivative of the projection with respect to the coordinates is zero or in mathematical terminology is invariant and

$$\frac{d}{ds}\left(A_u \frac{dx_u}{ds}\right) = 0$$

The problem is to find a mathematical solution for the metric tensor either by processing observed motion in an orbit determination program or by solving the above equations. The above computer approach requires immersion of a spacecraft in the gravity field which can be accomplished by passing an electromagnetic wave between two spacecraft safely away from the distributed mass. The analytic approach requires observation of the mass distribution and solution of differential equations that are really difficult to solve. It may be difficult, if not impossible, to observe the distribution of mass in a black hole. Performing the differentiation,

$$\frac{\partial A_u}{\partial x_v}\frac{dx_v}{ds}\frac{dx_u}{ds} + A_u \frac{d^2 x_u}{ds^2} = 0 \tag{1.7}$$

This result is of practical use if it is applied to a geodesic. From Eq. (1.3), we have

$$A_\alpha \frac{d^2 x_\alpha}{ds^2} = -A_\alpha \{uv, \alpha\} \frac{dx_u}{ds}\frac{dx_v}{ds}$$

Applying this result to Eq. (1.7) gives

$$\frac{dx_u}{ds}\frac{dx_v}{ds}\left(\frac{\partial A_u}{x_v} - A_\alpha \{uv, \alpha\}\right) = 0$$

The expression in the brackets is the covariant derivative of a vector and is given by

$$A_{i,j} = \frac{\partial A_i}{\partial x_j} - \{ij, \alpha\} A_\alpha = 0 \tag{1.8}$$

The covariant derivative is the x_jth derivative of A_i with respect to the metric tensor g_{ij}. The covariant derivative cannot be solved for the metric tensor because of the presence of the vector A so A must be eliminated. One more differentiation is needed. In the classical world we differentiate position to get velocity and then differentiate velocity to get acceleration. We then write equations for acceleration and then integrate twice to get position. We do the same thing in the curved space world. The location of mass in the real world defines the curvature of space or the metric tensor. The metric tensor gives us the equations of motion through the equation of geodesics.

1.6   Covariant Derivative of a Tensor

Consider the following product of a covariant tensor with a contravariant vector.

$$A_{uv} \frac{dx_u}{ds} \frac{dx_v}{ds}$$

A_{uv} is the covariant derivative defined by Eq. (1.8). The first product of the x_u term with the rows of A_u defines a vector. The second product of the x_v term results in a scalar dot product. Differentiating with respect to the coordinates as was done for the covariant vector, we obtain

$$\frac{\partial A_{uv}}{\partial x_\sigma} \frac{dx_\sigma}{ds} \frac{dx_u}{ds} \frac{dx_v}{ds} + A_{uv} \frac{dx_v}{ds} \frac{d^2 x_u}{ds^2} + A_{uv} \frac{dx_u}{ds} \frac{d^2 x_v}{ds^2} = 0$$

Substituting Eq. (1.8), we obtain the covariant derivative of a tensor.

$$\frac{d^2 x_u}{ds^2} = -\{u\sigma, \alpha\} \frac{dx_u}{ds} \frac{d\sigma}{ds}$$

$$\frac{d^2 x_v}{ds^2} = -\{v\sigma, \alpha\} \frac{dx_v}{ds} \frac{d\sigma}{ds}$$

$$\left(\frac{\partial A_{uv}}{\partial x_\sigma} - A_{\sigma v} \{u\sigma, \alpha\} - A_{u\sigma} \{v\sigma, \alpha\} \right) \frac{dx_\sigma}{ds} \frac{dx_u}{ds} \frac{dx_v}{ds} = 0$$

$$A_{uv\sigma} = \frac{\partial A_{uv}}{\partial x_\sigma} - A_{\sigma v} \{u\sigma, \alpha\} - A_{u\sigma} \{v\sigma, \alpha\}$$

or

$$A_{i,jk} = \frac{\partial A_{i,j}}{\partial x_k} - \{ik, \alpha\} A_{\alpha,j} - \{jk, \alpha\} A_{i,\alpha} \tag{1.9}$$

1.7 Riemann–Christoffel Tensor

Substituting the covariant derivative of a vector into the covariant derivative of a tensor, we obtain

$$
A_{i,jk} = \frac{\partial}{x_k} \left(\frac{\partial A_i}{\partial x_j} - \{ij, \alpha\} A_\alpha \right) - \{ik, \alpha\} \left(\frac{\partial A_\alpha}{\partial x_j} - \{\alpha j, \beta\} A_\beta \right)
$$

$$
- \{jk, \alpha\} \left(\frac{\partial A_i}{\partial x_\alpha} - \{i\alpha, \gamma\} A_\gamma \right) \tag{1.10}
$$

If we reverse the order of differentiation, we get

$$
A_{i,kj} = \frac{\partial}{x_j} \left(\frac{\partial A_i}{\partial x_k} - \{ik, \alpha\} A_\alpha \right) - \{ij, \alpha\} \left(\frac{\partial A_\alpha}{\partial x_k} - \{\alpha k, \beta\} A_\beta \right)
$$

$$
- \{kj, \alpha\} \left(\frac{\partial A_i}{\partial x_\alpha} - \{i\alpha, \gamma\} A_\gamma \right) \tag{1.11}
$$

Carrying out the differentiation

$$
A_{i,jk} = \frac{\partial^2 A_i}{\partial x_k x_j} - \frac{\partial \{ij, \alpha\}}{\partial x_k} A_\alpha - \{ij, \alpha\} \frac{\partial A_\alpha}{\partial x_k} - \{ik, \alpha\} \frac{\partial A_\alpha}{x_j}
$$

$$
+ \{ik, \alpha\} \{\alpha j, \beta\} A_\beta - \{jk, \alpha\} \frac{\partial A_i}{\partial x_\alpha} - \{jk, \alpha\} \{i\alpha, \gamma\} A_\gamma
$$

$$
A_{i,kj} = \frac{\partial^2 A_i}{\partial x_j x_k} - \frac{\partial \{ik, \alpha\}}{\partial x_j} A_\alpha - \{ik, \alpha\} \frac{\partial A_\alpha}{\partial x_j} - \{ij, \alpha\} \frac{\partial A_\alpha}{x_k}
$$

$$
+ \{ij, \alpha\} \{\alpha k, \beta\} A_\beta - \{kj, \alpha\} \frac{\partial A_i}{\partial x_\alpha} - \{kj, \alpha\} \{i\alpha, \gamma\} A_\gamma
$$

The order of differentiation should not make any difference so, if we subtract, the result should be zero. We are looking for a metric tensor that gives this result.

$$
A_{i,jk} - A_{i,kj} = \{ik, \alpha\} \{\alpha j, \beta\} A_\beta - \frac{\partial \{ij, \alpha\}}{\partial x_k} A_\alpha - \{ij, \alpha\} \{\alpha k, \beta\} A_\beta
$$

$$
+ \frac{\partial \{ik, \alpha\}}{\partial x_j} A_\alpha = 0
$$

Interchanging the α and β dummy indexes associated with the A_β terms and factoring out A_α we obtain the Riemann–Christoffel tensor,

$$
R_{ijk}^\alpha = \{ik, \beta\} \{\beta j, \alpha\} - \frac{\partial \{ij, \alpha\}}{\partial x_k} - \{ij, \beta\} \{\beta k, \alpha\} + \frac{\partial \{ik, \alpha\}}{\partial x_j} = 0 \tag{1.12}
$$

This tensor has the property we need to solve for the metric tensor. The arbitrary measurement vector A has been eliminated and the Christoffel symbols are a function of only the metric tensor.

The Riemann–Christoffel tensor has 256 elements and each element is a function of the metric tensor. We would have to solve 256 simultaneous differential equations to obtain a solution. Due to symmetry, most of the elements of the Riemann–Christoffel tensor are equal to or multiples of a subset of independent elements. All the elements of the Riemann–Christoffel tensor are zero if the independent elements are zero. In order to isolate a set of independent elements, a number of identities associated with the symmetry of the metric tensor and Christoffel symbols are defined. From Eq. (1.12) it is immediately obvious that the Riemann tensor is skew symmetric on the last two indices.

$$R^i_{jkl} = -R^i_{jlk}$$

We define the covariant form of the Riemann tensor by

$$R_{ijkl} = g_{i\alpha} R^\alpha_{jkl}$$

Multiplying through by $g^{i\beta}$ we obtain

$$R^\beta_{jkl} = g^{i\beta} R_{ijkl}$$

Substituting the Christoffel symbols

$$R_{ijkl} = \frac{1}{2} \left(\frac{\partial^2 g_{il}}{\partial x^j x^k} - \frac{\partial^2 g_{jl}}{\partial x^i x^k} - \frac{\partial^2 g_{ik}}{\partial x^j x^l} + \frac{\partial^2 g_{jk}}{\partial x^i x^l} \right)$$

$$+ g^{\alpha\beta} \left([jk, \beta][il, \alpha] - [jl, \beta][ik, \alpha] \right)$$

where

$$\{ij, k\} = g^{k\alpha} [ij, \alpha]$$

$$[ij, \alpha] = \frac{1}{2} \left(\frac{\partial g_{ik}}{\partial x^j} + \frac{\partial g_{jk}}{\partial x^i} - \frac{\partial g_{ij}}{\partial x^k} \right)$$

and it is obvious that

$$R_{ji,kl} = -R_{ijkl}$$

$$R_{ij,lk} = -R_{ijkl}$$

$$R_{ji,kl} = R_{ijkl}$$

$$R_{ijkl} + R_{iklj} + R_{iljk} = 0$$

and after raising the index

$$R^i_{jkl} + R^i_{klj} + R^i_{ljk} = 0$$

1.8   Ricci Tensor

In tensor algebra, the independent elements of the Riemann tensor can be isolated by what is called contraction. In Einstein summation notation, contraction is performed by making the k and α indices of Eq. (1.12) equal. The result is the Ricci tensor.

$$R_{ij} = \{i\alpha, \beta\}\{\beta j, \alpha\} - \frac{\partial \{ij, \alpha\}}{\partial x_\alpha} - \{ij, \beta\}\{\beta\alpha, \alpha\} + \frac{\partial \{i\alpha, \alpha\}}{\partial x_j} \qquad (1.13)$$

The R symbol stands for Riemann, not Ricci. Einstein does not mention Ricci when he contracts the Riemann tensor. He just makes the first and third index equal and calls it contraction. Einstein does mention Ricci in his tutorial on tensors. The Ricci tensor is solved for the metric tensor and when the boundary condition at infinity is applied the equations of motion are obtained.

The simple operation of contraction which results in the Ricci tensor must have some mathematical properties that enable it to be solved for the metric tensor. The Ricci tensor must be symmetrical and equal to zero. We could guess a solution for the metric tensor and insert it into the Riemann tensor. If all 256 elements are zero, we have a solution. Since most of the elements of the Riemann tensor are equal to other elements, we can extract a subset of the Riemann tensor elements and solve these for the metric tensor. If one element of the Riemann tensor is non zero, then at least one element of the reduced subset is non zero. Conversely, if all the elements of the reduced subset are zero, all the elements of the Riemann tensor are zero. It can be shown that the Ricci contraction has this property.

The Ricci tensor has 16 elements. There are 10 unknown elements in the metric tensor. We need 10 independent equations to solve for the metric tensor. There are 10 independent equations in the Ricci tensor if it is symmetrical. The proof of symmetry is not trivial. We start with the determinant of the metric tensor

$$g = |g_{ij}|$$

$$g = \sum_{j=1}^{4} g_{ij} G^{ij}$$

$$g = g_{ij} \, G^{ij} \text{ (sum on } j \text{ only, } i \text{ fixed)}$$

$$gg^{ij} = G^{ij} \tag{1.14}$$

G are the minors of g_{ij}. Differentiating the determinant g with respect to g_{ij} gives

$$\frac{\partial g}{\partial g_{ij}} = g_{i\alpha} \frac{\partial G^{i\alpha}}{\partial g_{ij}} + G^{i\alpha} \frac{\partial g_{i\alpha}}{\partial g_{ij}}$$

Since the minor of g_{ij} is not a function of g_{ij} we have

$$\frac{\partial g}{\partial g_{ij}} = G^{i\alpha} \frac{\partial g_{i\alpha}}{\partial g_{ij}} = G^{i\alpha} \delta_\alpha^j = G^{ij}$$

The partial derivative of $g_{i\alpha}$ with respect to g_{ij} is one when $j = \alpha$ and zero when $j \neq \alpha$. The partial derivative of something with respect to itself is one and the partial derivative of something with respect to something else is zero provided something else is independent. But

$$\frac{\partial g}{\partial x_i} = \frac{\partial g}{\partial g_{\alpha\beta}} \frac{\partial g_{\alpha\beta}}{\partial x_i} = G^{\alpha\beta} \frac{\partial g_{\alpha\beta}}{\partial x_i}$$

and since from Eq. (1.14) $G^{\alpha\beta} = gg^{\alpha\beta}$ we get

$$\frac{\partial g}{\partial x_i} = gg^{\alpha\beta} \frac{\partial g_{\alpha\beta}}{\partial x_i} \tag{1.15}$$

From the definition of the Christoffel symbols we get

$$[\alpha i, \gamma] = \frac{1}{2} \left(\frac{\partial g_{\alpha\gamma}}{\delta x_i} + \frac{\partial g_{i\gamma}}{\delta x_\alpha} - \frac{\partial g_{\alpha i}}{\delta x_\gamma} \right)$$

$$[\gamma i, \alpha] = \frac{1}{2} \left(\frac{\partial g_{\gamma\alpha}}{\delta x_i} + \frac{\partial g_{i\alpha}}{\delta x_\gamma} - \frac{\partial g_{\gamma i}}{\delta x_\alpha} \right)$$

Adding we get

$$\frac{\partial g_{\alpha\gamma}}{\delta x_i} = [\alpha i, \gamma] + [\gamma i, \alpha]$$

Inserting this result into Eq. (1.15) we get

$$\frac{\partial g}{\partial x_i} = gg^{\alpha\beta} \left(g_{\gamma\beta} \{\alpha i, \gamma\} + g_{\gamma\alpha} \{\beta i, \gamma\} \right)$$

$$\frac{\partial g}{\partial x_i} = g\left(\{\alpha i, \alpha\} + \{\beta i, \beta\}\right)$$

$$\frac{\partial g}{\partial x_i} = 2g\{\alpha i, \alpha\}$$

$$\frac{1}{2g}\frac{\partial g}{\partial x_i} = \{i\alpha, \alpha\}$$

$$\frac{\partial}{\partial x^i}\log\sqrt{g} = \{i\alpha, \alpha\} \tag{1.16}$$

$$R_{i,j} = \frac{\partial^2 \log\sqrt{g}}{\partial x^j \partial x^i} - \frac{\partial \{ij, \alpha\}}{\partial x^\alpha} + \{i\alpha, \beta\}\{\beta j, \alpha\} - \{ij, \beta\}\frac{\partial \log\sqrt{g}}{\partial x^\beta}$$

The first term of $R_{i,j}$ is symmetrical because the order of differentiation is arbitrary. If we interchange the i and j indices, we get the same rank two tensor. The second term is symmetric because the Christoffel symbols are symmetric on the first two indices and therefore the rank two tensor formed by contraction on α is also symmetric. The Christoffel symbols from Eq. (1.4) are given by

$$\{uv, \alpha\} = \frac{1}{2}g^{\sigma\alpha}\left(\frac{\partial g_{u\sigma}}{\delta x_v} + \frac{\partial g_{v\sigma}}{\delta x_u} - \frac{\partial g_{uv}}{\delta x_\sigma}\right) \tag{1.17}$$

If we interchange the u and v indices, the first and second terms in the brackets interchange positions and the sum is the same. The third term is symmetrical because the metric tensor is symmetrical. The fourth term is symmetrical when we contract on β

The reader can skip to Chap. 2 since we do not need the Einstein tensor to do navigation. We are only interested in spherical symmetrical empty space where spacecraft go. The constant of integration can be obtained by applying Newton's law of gravity as a boundary condition at infinity. The mathematics become a lot easier from Chap. 2 to the end where an experiment is performed to verify Einstein and the computer program used for navigation. The latter is the primary objective of this book.

1.9   Einstein Tensor

The Ricci tensor is zero where there is no mass. We can neglect the acceleration inside the sun, or at black holes, which are not relevant to current spacecraft trajectories. Outside the sun, we can force the Ricci tensor to satisfy Newton's law of gravity for the case of spherical symmetry. A problem with the Ricci tensor is that it models the curvature of space but does not account for the scale. This problem with formulating measurements is common when determining orbits. For example, when tracking landmarks to determine an orbit about an asteroid, the angle measurements cannot determine the scale. Angles cannot determine length. Doppler or range data must be introduced to determine length. Another example is determining the orbit of planets by measuring angles obtained from a photographic plate on a star background. The astronomical unit or distance of the Earth from the Sun must be obtained from other sources of data. The astronomical unit was originally determined by observing the orbit of Eros near the Earth. A more relevant example is determining the inertia tensor of a rotating body by observation of its rotation. The trace of the inertia tensor can only be determined if some known external torque is applied to the body. The Einstein tensor acknowledges this problem and adds a term to the Ricci tensor.

The Einstein tensor applies both inside and outside the sun. Therefore the challenge for Einstein was to find a term that when added to the Ricci tensor satisfies the boundary condition at the surface of the sun and scales the Ricci tensor. We could, as suggested by Eddington, just multiply the Ricci tensor by a constant proportional to density. Eddington points out that this approach does not work. Since the volume elements in curved space vary in size, constant density would imply variable mass. Since energy is mass and mass creates space, conservation of energy requires the divergence of space, or increase in energy, to be zero in free space and equal to mass where mass is present. For a volume integral to give the correct volume, all the volume elements must be independent. In curved space, the size and shape of all the volume elements change every time we add mass. Einstein invented a term called invariant density to get around this problem. His approach was to map the Riemann tensor to Euclidean coordinates and define regular volume elements of equal size and equal mass for constant density. He then contracts the Riemann tensor in Euclidean space to form the Einstein tensor and maps back to curved space thus conserving mass.

If we lower the index i on the contravariant Riemann tensor and differentiate with respect to x_m we get

$$R_{ijkl,m} = g_{i\alpha} R^{\alpha}_{jkl,m} = \frac{\partial^2 \{jl, i\}}{\partial x_m \partial x_k} - \frac{\partial^2 \{jk, i\}}{\partial x_m \partial x_l} \tag{1.18}$$

Lowering the index i on the Riemann tensor effectively maps the line element from curved space coordinates to Euclidean. Permuting the k, l, m indices and adding yields the Bianchi identities

$$R_{ijkl,m} + R_{ijlm,k} + R_{ijmk,l} = 0$$

Recall that the divergence is the second derivative of a potential function with respect to each of the coordinates summed. Permuting the indices and adding is equivalent to summing the flow of energy or space into and out of a volume element. If there is no mass present in the volume element, this result is zero. If there is mass present, the increase in the curvature of space is proportional to the mass. We then perform contraction by applying the appropriate skew-symmetric properties of the Riemann tensor and then multiplying by $g^{il}\, g^{jk}$ to map back to curved space coordinates. An abbreviated derivation is given by Sokolnikoff in Reference [4].

$$g^{jk}(g^{il} R_{ijkl,m}) + g^{jk}(g^{il} R_{ijlm,k}) + g^{il}(g^{jk} R_{ijmk,l}) = 0$$

The order of multiplication by the $g^{il}\, g^{jk}$ does not matter because they operate on different indices of the Riemann tensor. Next, we use the skew-symmetric properties of the Riemann tensor to obtain

$$g^{jk}(g^{il} R_{ijkl,m}) - g^{jk}(g^{il} R_{ijml,k}) - g^{il}(g^{jk} R_{jimk,l}) = 0$$

Contracting on the first and fourth index we get

$$g^{jk} R_{jk,m} - g^{jk} R_{jm,k} - g^{il} R_{im,l} = 0$$

and

$$g^{ij} R_{ij,m} - 2R^k_{m,k} = 0$$

or

$$\left(R^k_m - \frac{1}{2}\delta^k_m R \right)_{,k} = 0 \qquad (1.19)$$

$$R = g^{ij} R_{ij}$$

Eq. (1.19) states that the divergence of space or the Einstein tensor is zero. The Einstein tensor is

$$G^i_j = R^i_j - \frac{1}{2}\delta^i_j R$$

or in the covariant form

$$G_{ij} = R_{ij} - \frac{1}{2}g_{ij} R \qquad (1.20)$$

The Einstein tensor is equated to what is called the stress–energy tensor

$$G_{ij} = \frac{8\pi G}{c^2} T_{ij} \qquad (1.21)$$

Outside the sun, **T** is equal to zero. Inside the sun

$$
T_{ij} = \begin{bmatrix} p & 0 & 0 & 0 \\ 0 & p & 0 & 0 \\ 0 & 0 & p & 0 \\ 0 & 0 & 0 & \rho \end{bmatrix}
\tag{1.22}
$$

The variable ρ is the density of matter and the variable p is the pressure differential that is obtained in hydrostatic equilibrium.

Equating G to T is not a mathematically derived result but a statement of physics that cannot be proven. The reader may wonder why Einstein did not just equate the Ricci tensor with T. Both the Ricci tensor and Einstein tensor are curvature tensors. Take the shape of the sun or any body and move it to flat Euclidean space. Partition the body into a large number (infinity) of cubes. Assuming constant density, calculate the mass of each cube. Map the cubical volume elements to curved space and fill each curved space volume elements with the mass computed for each cubical volume element. This operation conserves the mass. We now have a problem because the curved space volume elements have different volumes and densities. For flat space, the volume and mass can be computed by adding the volume elements in any order. For curved space coordinates the mass sum of all the volume elements will depend on the order if we assume constant density. If we build up the sun one volume element at a time, the mass will vary because each volume element changes the curved space coordinates. The Einstein tensor solves this problem by book keeping the mass in Euclidean space and mapping to curved space.

1.10   Summary of Einstein's Theory

Before solving the field equations for a specific metric tensor and applying this result to the equation of geodesics to obtain the equations of motion, a review of Einstein's and classical potential theory may be useful. Comparison of the general relativity approach with Newton's classical theory reveals some striking similarities. Newton starts with his inverse square law and the potential and divergence follow in a straight forward mathematical derivation. The divergence is solved for the potential, and the equations of motion follow from the gradient of the potential function. The inverse square law is given and is not proven.

Einstein starts from a much simpler equation and after a much more difficult mathematical derivation obtains the Einstein tensor which can be solved for the metric tensor. The simple equation he starts from is difficult to interpret in the physical world. Eddington has no problem in not understanding in that he argues that if the resulting equations of motion can be verified by experiment, the theory must be correct. Einstein offered a physical explanation of the simple starting point in terms of tensors. The difference between Einstein and Eddington is minor and in the

Table 1.1 Comparison of Newton's theory with Einstein's theory

Newton's theory			Einstein's theory
$\sum_{i=0}^{\infty} \dfrac{-\mu_i}{\lvert \mathbf{r}' - \mathbf{r}_i \rvert}$	U		$A_u \dfrac{dx_u}{ds}$
$\sum_{i=0}^{\infty} \dfrac{-\mu_i(\mathbf{r}' - \mathbf{r}_i)}{\lvert \mathbf{r}' - \mathbf{r}_i \rvert^3}$	∇U		$\dfrac{\partial A_i}{\partial x_j} - \{ij, \alpha\} A_\alpha$
$\sum_{i=0}^{\infty} \dfrac{3\mu_i}{\lvert \mathbf{r}' - \mathbf{r}_i \rvert^3} - \dfrac{3\mu_i}{\lvert \mathbf{r}' - \mathbf{r}_i \rvert^5} \lvert \mathbf{r}' - \mathbf{r}_i \rvert^2 = 0$	$\nabla \cdot \nabla U = \nabla^2 U$		$R_{uv} - \tfrac{1}{2} g_{uv} R$

end they agree. Eddington probably preferred the experimental approach since he conducted the light bending experiment that proved the theory. Einstein provides a verbal description of general relativity and then defines certain properties of tensors. He then presented the mathematics of general relativity and used the properties of tensors to prove the equations. Sometimes the physics is difficult to discern, but Einstein was more interested in convincing mathematicians than physicists. It appears Einstein was trying to convince mathematicians that he was one of them and Eddington, who was a mathematician, was trying to convince physicists that he was one of them. Einstein's theory of general relativity must meet a stringent requirement. If one step in the derivation is in error, the whole theory is lost.

Table 1.1 contains the key mile stones in the development of Newton's classical theory of gravity and Einstein's General Theory of Relativity. For the classical theory, the first row contains a scalar potential, the second row contains the first derivative or acceleration of the scalar potential, and the third row contains the second derivative or divergence of the scalar potential. For general relativity, the first row contains a scalar which may be thought of as a measurement, the second row contains the first derivative or covariant derivative of the measurement, and the third row contains the second derivative or curvature of the measurement which is the Einstein tensor. For Newton's theory, the inverse square law is given and the potential function and divergence may be obtained by mathematical operations on the inverse square law. We simply integrate to obtain the potential and differentiate to obtain the divergence. It is not necessary to assign any physical meaning to potential or divergence. For general relativity, the term in the first row is an artificial measure of motion when applied to the equation of geodesics. The covariant derivative and Einstein tensor are obtained by mathematical operations and require no physical explanation.

The key to physically understanding theories involving the divergence theorem is to physically describe one of the terms given in Table 1.1. The other two terms become locked in once one is understood since they are mathematically related. For classical gravitational theory, the potential and divergence are tied to the inverse square law and we can solve the equation for divergence and take the gradient to obtain the equations of motion. For heat, the divergence can be obtained directly from heat flow and solved for the scalar temperature distribution.

For general relativity, we define a measurement that consists of the product of two vectors.

$$Z = A_u \frac{dx_u}{ds}$$

A_u can be thought of as acceleration and $\frac{dx_u}{ds}$ is in the direction of the velocity vector or tangent to the line element at some point. Both of these vectors are observable but their components are dependent on the assumed coordinate system. The dot product (Z) is not dependent on the assumed coordinate system. The projection of A_u on $\frac{dx_u}{ds}$ is the same in any coordinate system. Z is said to be invariant with respect to the system of coordinates. The derivative of Z with respect to the coordinates (the covariant derivative) is also invariant as are higher order derivatives. The solution of the resulting differential equations is obtained by applying insight into the constraints associated with symmetry. One can go through all the mathematics associated with this solution and have no understanding of general relativity. The understanding of General Relativity theory is associated with the above fundamental assumptions and the geometry which is subsumed by the mathematics.

The derivation and proof that the Einstein tensor is equal to curvature which is equated to the stress–energy tensor is probably the most difficult part of General Relativity Theory to understand. Even Sir Arthur Eddington had trouble understanding Einstein's tensor and he was one of the three people in the world who understood general relativity. Remember the joke when Eddington replied "who was the third?". Consider the following quote from page 115 of Eddington's book (Reference [2]).

The divergence of $G^i_{\ j} = R^i_{\ j} - \frac{1}{2}\delta^i_{\ j} R$ is identically zero (52)

"I think it should be possible to prove (52) by geometrical reasoning … But I have not been able to construct a geometrical proof and must content myself with a clumsy analytic verification."

What is needed is a clumsy geometrical verification. The following geometrical derivation lacks the rigor of the mathematical derivation. Because any error invalidates the entire theory, geometrical descriptions are often shunned in the literature. The reader should regard the following geometrical derivation as suspicious. It is provided because the authors work with matrices and are not that familiar with summation notation.

Geometry is best described by vectors. Let the curved space coordinate $(\mathbf{x_c})$ be defined by the coordinate transformation T from Euclidean coordinates (x_e).

$$\mathbf{x_c} = T\ \mathbf{x_e} \tag{1.23}$$

and

$$\mathbf{x_e} = T^{-1}\ \mathbf{x_c} \tag{1.24}$$

The matrix T is said to be covariant and its inverse (T^{-1}) is contravariant. In flat space T is an orthogonal transformation matrix and in curved space T can be anything depending on the universe. Our universe imposes some restrictions on T associated with symmetry and the geometry, that is that there exists a one to one mapping from Euclidean space to curved space. In flat space $\mathbf{x}_e$ and $\mathbf{x}_c$ are the same. The mapping from curved space to Euclidean space is obtained by inverting T. The transformation from Euclidean space to curved space is not unique and T may be complex depending on the definition of the coordinates.

The line element is defined by

$$ds_c^2 = \mathbf{x}_\mathbf{c}^T \, \mathbf{x}_\mathbf{c} = (T \, \mathbf{x}_\mathbf{e})^T \, (T \, \mathbf{x}_\mathbf{e}) = \mathbf{x}_\mathbf{e}^T \, T^T \, T \, \mathbf{x}_\mathbf{e}$$

The matrix defined by $T^T \, T$ is called the metric tensor g_{uv} and we have

$$ds_c^2 = \mathbf{x}_e^T \, g_{uv} \, \mathbf{x}_e = \mathbf{x}_c^T \, I \, \mathbf{x}_c \tag{1.25}$$

and after multiplying by the inverse of T we have

$$ds_e^2 = \mathbf{x}_e^T \, I \, \mathbf{x}_e = \mathbf{x}_c^T \, g_{uv}^{-1} \, \mathbf{x}_c \tag{1.26}$$

For an orthogonal transformation in flat space g_{uv} is positive definite. A transformation matrix is effectively the square root of g_{uv}. Because of the minus sign in the Minkowski metric, T is complex. Einstein got around complex transformation matrices by defining the time coordinate as a complex number (ict).

In curved space the metric tensor is not constant but varies as we move away from a point of interest. The above equation for the line element is thus only valid for infinitely small $\mathbf{x}$. The vectors $\mathbf{x}_e$ and $\mathbf{x}_c$ from the origin to a point in space are therefore not useful. In flat space, ds can be of any length. Vectors are straight lines. In curved space we must integrate to define a line. Therefore, the vectors are only useful as differentials.

In the derivation of the Einstein tensor there are frequent multiplications by g_{uv} or g^{uv}. It may be useful to attempt to do these operations with matrices. The Ricci tensor may be diagonalized into a transformation matrix by taking the square root. Thus we have

$$R = \sqrt{R}^T \sqrt{R}$$

and we way transform a vector from Euclidean space to curved space by

$$Y_c'' = \sqrt{R} X_e$$

The second derivative of the curved space line element is thus

$$ds_c^{2''} = Y_c''^T \, Y_c'' = X_e^T \sqrt{R}^T \sqrt{R} X_e$$

or

$$ds_c^{2''} = Y_c''^T Y_c'' = X_e^T R_{uv} X_e = 0 \tag{1.27}$$

The first derivative $(ds_c^{2'})$ must also be zero. In general, if the second derivative is zero, the first derivative is a maximum, a minimum, or an inflection. An inflection is a constant which may not be zero. Since the Riemann tensor, defined by the difference between Eqs. (1.10) and (1.11), is zero the first derivative or covariant derivative is also zero. The Christoffel symbols that multiply the covariant derivative are not zero. We can transform from curved space coordinates back to Euclidean coordinates by

$$Y_e'' = \sqrt{g^{-1}}\sqrt{R} X_e$$

The Euclidean line element is now

$$ds_e^{2''} = Y_e''^T Y_e'' = X_e^T \left(\sqrt{g^{-1}R}\right)^T \left(\sqrt{g^{-1}R}\right) X_e$$

or

$$ds_e^{2''} = Y_e''^T Y_e'' = X_e^T (g^{uv} R_{uv}) X_e \tag{1.28}$$

Equation (1.27) is the transformation from Euclidean space to the line element in curved space. The Euclidean coordinates are needed to define the volume elements such that the density is constant and the divergence of space through the faces is accounted for correctly. Equation (1.28) is the transformation from curved space to Euclidean. The curved space coordinates are needed for the equations of motion and for equating curvature to mass.

Transformations between Euclidean coordinates and curved space coordinates are achieved by multiplication by the covariant or contravariant metric tensor. Mathematicians and Einstein need not be concerned with the coordinates since they operate on tensors to define differential equations that can be solved for the metric tensor. Once the metric tensor is obtained, it is inserted into the equation of geodesics where the curved space coordinates are of interest. Summation notation defines the order of multiplication by the order of the indices. A subscripted index indicates a column and a superscripted index indicates a row for the matrices defined above. An index can appear no more than two times in a term of an equation indicating how the rows and columns are multiplied. When the index of a tensor is raised by multiplying through by the inverse of the metric tensor or contravariant metric tensor (g^{uv}), the mapping is to curved space coordinates. In summation notation the indices only apply to an equation and are only used to indicate the order of multiplication. A given equation can be rewritten with completely different indices since they have no physical meaning.

The Riemann tensor describes curvature at a point. It has no physical meaning because in the limit as dx approaches zero it vanishes. The Einstein tensor gives the Riemann tensor physical meaning. The contraction that yields the Ricci tensor is good for empty space but breaks down when there is mass present. In the literature the Einstein tensor is often referred to as an alternate contraction of the Riemann tensor. Here we think of it as a repair of the Ricci tensor. If the repair is not done correctly, the Einstein tensor will not equal the stress–energy tensor. The first step in Sect. 1.8 was to lower the index on the Riemann tensor. This operation transformed to Euclidean coordinates Eq. (1.25). Next, we permute the indices and sum three Riemann tensors to define the Bianchi identities. This operation is essentially the same as performed by the Δ^2 operator in classical theory. If we differentiate the Ricci tensor with respect to X_k and permute the indices we get

$$\frac{\partial R_{ij}}{\partial X_k} = \frac{\partial R_{ij}}{\partial X_1}, \frac{\partial R_{ij}}{\partial X_2}, \frac{\partial R_{ij}}{\partial X_3}, \frac{\partial R_{ij}}{\partial X_4}$$

$$\frac{\partial R_{ij}}{\partial X_l} = \frac{\partial R_{ij}}{\partial X_2}, \frac{\partial R_{ij}}{\partial X_3}, \frac{\partial R_{ij}}{\partial X_1}, \frac{\partial R_{ij}}{\partial X_4}$$

$$\frac{\partial R_{ij}}{\partial X_m} = \frac{\partial R_{ij}}{\partial X_3}, \frac{\partial R_{ij}}{\partial X_1}, \frac{\partial R_{ij}}{\partial X_2}, \frac{\partial R_{ij}}{\partial X_4}$$

The Bianchi identity is defined by

$$B_{ij} = \frac{\partial R_{ij}}{\partial X_k} + \frac{\partial R_{ij}}{\partial X_l} + \frac{\partial R_{ij}}{\partial X_m} = 0$$

For spherical symmetry, B_{ij} is diagonal and the diagonal elements of B_{ij} are given by

$$B_{11} = \frac{\partial R_{11}}{\partial X_1} + \frac{\partial R_{22}}{\partial X_2} + \frac{\partial R_{33}}{\partial X_3} = B_{22} = B_{33}$$

$$B_{44} = 3 \frac{\partial R_{44}}{\partial X_4}$$

The first three rows of B correspond to the mass associated with the volume element. Since

$$\frac{\partial B_{ij}}{\partial X_k} = 0$$

$$\frac{\partial R_{11}}{\partial X_1} + \frac{\partial R_{22}}{\partial X_2} + \frac{\partial R_{33}}{\partial X_3} - \frac{\partial R_{44}}{\partial X_4} = 0 \qquad (1.29)$$

and

$$\frac{\partial R_{44}}{\partial \mathbf{X}_4} = \frac{\partial R_{11}}{\partial \mathbf{X}_1} + \frac{\partial R_{22}}{\partial \mathbf{X}_2} + \frac{\partial R_{33}}{\partial \mathbf{X}_3} \tag{1.30}$$

The B_{44} diagonal element may be equated with the three spatial coordinates. First, we have to deal with the three in the B_{44} equation. Since the Bianchi identity adds up three Ricci tensors, we have to divide by three to get the correct scale. That requires that the spatial terms (B_{11}, B_{22}, B_{33}) be divided by three. Since the stress–energy tensor volume element is projected along the direction of the gravity field, we must divide by three. Multiplying the spatial terms by three scales both sides of the equations.

Now we can factor out the diagonal elements of B_{ij} which are the trace of R_{ij}

$$B_{ij} = \frac{\partial I R}{\partial \mathbf{X}_k}$$

$$R = \frac{\partial R_{11}}{\partial \mathbf{X}_1} + \frac{\partial R_{22}}{\partial \mathbf{X}_2} + \frac{\partial R_{33}}{\partial \mathbf{X}_3} + \frac{\partial R_{44}}{\partial \mathbf{X}_4}$$

and I is the four by four identity matrix. R is a scalar when mapped to the line element in curved space. Since B_{ij} is effectively the local curvature of space, it can be equated with the stress–energy tensor T_{ij}. First, we must transform to curved space coordinates by raising the index on T_{ij}. Mass equals curvature in curved space.

$$\frac{\partial I R}{\partial \mathbf{X}_k} = k \frac{\partial T^i_j}{\partial \mathbf{X}_k}$$

If we start from empty space, R_{ij} is initially the Minkowski metric in the limit as m approaches zero. From the definition of the derivative we have the following difference equation.

$$R^{i+}_j = R^{i-}_j + \frac{\partial \Delta R^{i-}_j}{\partial X_k} \Delta X_k$$

where X_k are the coordinates of a volume element. The volume element is added on the surface of the body. The boundary condition requires that the Ricci tensor of the body at this point is equal to the Ricci tensor of free space. Equivalence of curvature and mass results in

$$\Delta R^i_j = \Delta(I R)$$

or

$$\Delta R^i_j = \Delta(k T^i_j)$$

Adding we get

$$2\Delta R^i_j = \Delta I R + \Delta k T^i_j$$

$$\Delta\left(R^i_j - \frac{1}{2}I R\right) = \Delta\frac{k}{2}T^i_j$$

The Einstein tensor is thus

$$G^i_j = R^i_j - \frac{1}{2}I R = \frac{k}{2}T^i_j$$

or in Euclidean coordinates

$$G_{ij} = R_{ij} - \frac{1}{2}g_{ij} R = 8\pi\,\frac{G}{c^2}T_{ij}$$

The constant k scales the stress–energy tensor to equal mass.

Let Pi be the ratio of the measured circumference of a circle to the diameter. In the above equation π is a constant defined by a series that happens to equal Pi in Euclidean coordinates. In curved space Pi is NOT equal to π. This will be verified by experiment later in Chap. 4. Einstein states in Reference [1] that PI is greater than π in curved space coordinates. Pi is actually a variable. Therefore, T_{ij} must be in Euclidean coordinates.

The pressure terms in the stress–energy tensor suggest the energy involved in compressing the matter within a volume element. The energy of compression, like associated with a spring, is included in the mass. Imagine a neutrally buoyant object like a jelly fish floating in the ocean. It is buoyed up by a force equal to the weight of the water displaced which is the weight of the jelly fish and is said to be in hydrostatic equilibrium. The integral of the pressure over the surface projected on to the gravity vector is equal to the weight or force of gravity. Now imagine a boulder buried deep in the Earth. It is also in hydrostatic equilibrium. We know this because the boulder is not moving. The excess pressure over the bottom of the boulder compared to the top times the projected area is equal to the weight of the boulder. The pressure differential is the pressure in the stress–energy tensor. It can be shown that the sum of the pressures associated with each spatial coordinate times the projected area is equal to the density times the volume. Thus, for a spherical symmetrical body like the sun, T_{22} and T_{33} in spherical coordinates are equal to zero and T_{11} is in the radial direction. The pressure (p) is numerically equal to T_{44} the density (ρ). Pressure and density have the same units (Newtons per square meter) in curved space when the time coordinate is normalized to meters. For the Schwarzschild solution in Chap. 2, the mass is determined from the G_{44} term of the Einstein tensor. The same mass could have been determined from the G_{11} term. But only if $p = \rho$.

An interesting aside may be deduced from Eq. (1.27). The elements of the Ricci tensor can be replaced by the gradient of a potential function. The scalar potential is curvature as defined by the line element ds. If we define G_w as

$$G_w = \frac{\partial \, ds^2}{\partial X}$$

Then

$$R_{ij} = \frac{\partial^2 ds^2}{\partial X^2} = \frac{\partial G_w}{\partial X}$$

Substituting into Eq. (1.29) gives

$$\frac{\partial^2 G_w}{\partial X_1^2} + \frac{\partial^2 G_w}{\partial X_2^2} + \frac{\partial^2 G_w}{\partial X_3^2} - \frac{\partial^2 G_w}{\partial X_4^2} = 0$$

If we replace X_1 by r and X_4 by ct, we have a wave equation propagating in the r direction.

$$\frac{\partial^2 G_w}{\partial r^2} - \frac{1}{c^2} \frac{\partial^2 G_w}{\partial t^2} = 0 \qquad (1.31)$$

Gravity waves are not of much interest to navigation. They are generally too small to be detected. However, if a substantial gravity wave propagated through the solar system it would perturb a spacecraft and the Earth and show up in the Doppler data residuals as two pulses. Two spacecraft along with the Earth would enable determination of the direction of the gravity wave. Unfortunately, a big gravity wave never came along. This may be fortunate because a really big gravity wave could shake the Earth's citizens off the planet. Gravity waves and radio waves share the same equation.

The author is well aware that the above derivation has some mathematical problems and obtaining the correct result does not justify the derivation. The intent here is to outline an approach that can eventually be turned into a proof by a mathematician. A problem with the above derivation is mixing Einstein summation notation with matrix notation. Introduction of matrix notation requires proof of some of the properties of matrices which may not be familiar to those familiar with summation notation. For example, the metric tensor is symmetrical because the product of any matrix with its transpose is symmetrical as shown in Eq. (1.25). The contraction of the Riemann tensor to obtain the Ricci tensor is achieved by equating two of the Riemann tensor indices. In matrix notation the Riemann tensor would be partitioned into 16 matrices that are of dimension 4 by 4 yielding a total of 256 elements. These matrices are obtained by differentiating the four matrices associated with the first derivative with respect to the four basis coordinates. The coordinate basis is factored out and the 16 matrices are summed to give the Ricci tensor. The proof of symmetery is simply the result that the derivatives of the metric tensor are symmetrical and the sum of 16 symmetrical matrices is also symmetrical.

In deriving the Einstein tensor, it was assumed that the Ricci tensor is diagonal. If the Ricci tensor is not diagonal, we can transform to a coordinate system where it is diagonal. This coordinate system would be the principal axes of the volume element. This is achieved by computing the eigenvectors and eigenvalues of the Ricci tensor and transforming coordinates using the matrix of eigenvectors as the orthoganal transformation matrix. A separate transformation matrix would be needed for every point in space or volume element. Since the trace of the Ricci tensor is equal to the sum of the eigenvalues, Einstein realized he could skip the transformation of coordinates and use the trace.

1.11   Summary

The Einstein field equations have been derived from fundamental principles as set forth by Einstein. In the remaining chapters, only the equation of geodesics and the Ricci tensor are needed to derive the equations used for navigation. The derivation of these equations is straight forward and may be found in the literature. Some physics has been ignored because their effect has been deemed to be insignificant. The sun is modeled as a rigid non rotating body in the shape of a sphere. The dragging of inertial frames from rotation of the sun is ignored for navigation. The perturbations of the planets are also ignored except for the determination of the planet ephemerides and the effect of Jupiter and Saturn gravity on clocks. The navigation software is initialized with a state vector obtained from a high precision planetary ephemeris which models the planets motion over many years. The integration of planet ephemerides for navigation is good for several months which is enough time for spacecraft orbit determination.

For the remaining chapters, the equations will be rigorously derived. Since these equations have been programmed into operational software, there is no room for error. The original implementation of these equations in the 1960s was by relativity experts at the California Institute of Technology and Jet Propulsion Laboratory who deserve all the credit. They will remain unnamed because the authors have not determined who they all are.

Chapter 2
Schwarzschild Solution for Spherical Symmetry

An exact solution of the Einstein field equations for a spherically symmetrical sun was obtained by Schwarzschild about a month after Einstein published his theory. Apparently Schwarzschild was aware of Einstein's work long before he published. It is unreasonable to believe that Schwarzschild was able to obtain his solution in a month. We know Hilbert was aware of Einstein's work because they were in communication with each other a few months before Einstein published his theory. According to Walter Isaacson in his book *Einstein*, Hilbert published his own theory, while Einstein was refining his theory. Einstein objected and Hilbert withdrew his paper. Hilbert recognized the theory was Einstein's and deferred. This little episode probably accelerated Einstein's publication. He probably would have preferred to wait a few more years, like Newton, because he knew that once the theory was published, he would have a lot of competition and distraction. He was right.

2.1 Schwarzschild Metric Tensor

A valid method for solving differential equations is to guess the solution, insert it into the equations, and hope it works. Schwarzschild started with a metric tensor that was obviously close to the actual metric but had a couple of undetermined functions.

$$g_{ij} = \begin{bmatrix} -e^\lambda & 0 & 0 & 0 \\ 0 & -r^2 & 0 & 0 \\ 0 & 0 & -r^2 \sin^2 \theta & 0 \\ 0 & 0 & 0 & e^\phi \end{bmatrix} \tag{2.1}$$

The zeros are due to symmetry. Since all the stars in the sky and the cosmic background are evenly distributed, it is reasonable to assume that there is no preferred direction for space. The only curvature is with respect to the r and time

© The Author(s), under exclusive license to Springer Nature Switzerland AG 2021
J. Miller, C. J. Weeks, *General Relativity for Planetary Navigation*, SpringerBriefs in Space Development, https://doi.org/10.1007/978-3-030-77546-9_2

spherical coordinates. The functions ϕ and λ were made exponents of e since he knew the g_{11} and g_{44} terms of the metric tensor would approach minus one and one, respectively (flat space) as r approached infinity. Once the metric tensor is defined, the Christoffel symbols may be computed from

$$\{uv, \alpha\} = \frac{1}{2} g^{\sigma\alpha} \left(\frac{\partial g_{u\sigma}}{\delta x_v} + \frac{\partial g_{v\sigma}}{\delta x_u} - \frac{\partial g_{uv}}{\delta x_\sigma} \right)$$

and

$$\{11, 1\} = \{rr, r\} = \frac{1}{2} \frac{d\lambda}{dr}$$

$$\{22, 1\} = \{\theta\theta, r\} = -re^{-\lambda}$$

$$\{33, 1\} = \{\phi\phi, r\} = -re^{-\lambda} \sin^2 \theta$$

$$\{44, 1\} = \{tt, r\} = \frac{1}{2} e^{\phi-\lambda} \frac{d\phi}{dr}$$

$$\{14, 4\} = \{rt, t\} = \frac{1}{2} \frac{d\phi}{dr}$$

$$\{13, 3\} = \{r\phi, \phi\} = \frac{1}{r}$$

$$\{23, 3\} = \{\theta\phi, \phi\} = \cot \theta$$

$$\{12, 2\} = \{r\theta, \theta\} = \frac{1}{r}$$

$$\{33, 2\} = \{\phi\phi, \theta\} = -\sin \theta \cos \theta \qquad (2.2)$$

The Christoffel symbols are inserted into the Ricci tensor. This is a tedious process and will be done for the R_{11} term. The other terms are easy to obtain once we have the R_{11} term.

$$R_{11} = \{1\alpha, \beta\} \{\beta\, 1, \alpha\} - \frac{\partial \{11, \alpha\}}{\partial x_\alpha} - \{11, \beta\} \{\beta\alpha, \alpha\} + \frac{\partial \{1\alpha, \alpha\}}{\partial x_1}$$

Term 1

$$\{1\alpha, \beta\} \{\beta\, 1, \alpha\} = \{11, 1\} \{11, 1\} + \{12, 2\} \{21, 2\} + \{13, 3\} \{31, 3\} + \{14, 4\} \{41, 4\}$$

$$\{11, 1\} \{11, 1\} = \frac{1}{4} (\lambda')^2$$

$$\{12, 2\} \{21, 2\} = \frac{1}{r^2}$$

$$\{13, 3\}\{31, 3\} = \frac{1}{r^2}$$

$$\{14, 4\}\{41, 4\} = \frac{1}{4}(\phi')^2$$

Term 2

$$\frac{\partial\{11, \alpha\}}{\partial x_\alpha} = \frac{\partial\{11, 1\}}{\partial x_1} = \frac{1}{2}\lambda''$$

Term 3

$$\{11, \beta\}\{\beta\alpha, \alpha\} = \{11, 1\}(\{11, 1\} + \{12, 2\} + \{13, 3\} + \{14, 4\})$$

$$\{11, \beta\}\{\beta\alpha, \alpha\} = \frac{1}{2}\lambda'\left(\frac{1}{2}\lambda' + \frac{1}{r} + \frac{1}{r} + \frac{1}{2}\phi'\right)$$

$$\{11, \beta\}\{\beta\alpha, \alpha\} = \left(\frac{1}{4}(\lambda')^2 + \frac{\lambda'}{r} + \frac{1}{4}\lambda'\phi'\right)$$

Term 4

$$\frac{\partial\{1\alpha, \alpha\}}{\partial x_1} = \frac{\partial}{\partial x_1}(\{11, 1\} + \{12, 2\} + \{13, 3\} + \{14, 4\})$$

$$= \frac{\partial}{\partial r}\left(\frac{1}{2}\lambda' + \frac{1}{r} + \frac{1}{r} + \frac{1}{2}\phi'\right)$$

$$+\frac{\partial\{1\alpha, \alpha\}}{\partial x_1} = \frac{1}{2}\lambda'' - \frac{2}{r^2} + \frac{1}{2}\phi''$$

The primes indicate differentiation with respect to r. The complete Schwarzschild–Ricci tensor is then given by

$$R_{11} = \frac{1}{2}\phi'' - \frac{1}{4}\lambda'\phi' + \frac{1}{4}(\phi')^2 - \frac{\lambda'}{r} = 0 \tag{2.3}$$

$$R_{22} = e^{-\lambda}\left[1 + \frac{1}{2}r(\phi' - \lambda')\right] - 1 = 0 \tag{2.4}$$

$$R_{33} = \sin^2\theta\left\{e^{-\lambda}\left[1 + \frac{1}{2}r(\phi' - \lambda')\right] - 1\right\} = 0 \tag{2.5}$$

$$R_{44} = e^{\phi-\lambda}\left[-\frac{1}{2}\phi'' + \frac{1}{4}\lambda'\phi' - \frac{1}{4}(\phi')^2 - \frac{\phi'}{r}\right] = 0 \tag{2.6}$$

The above equations are given on page 303 of Sokolnikoff's book [4] and on pages 330 and 331 of Lass's book [3]. We may conclude from Eqs. (2.3) and (2.6) that

$$\lambda' = -\phi'$$

and

$$\lambda = -\phi + \text{constant}$$

However, as r approaches infinity, λ and ϕ approach zero and the constant is also zero. Therefore,

$$\lambda = -\phi$$

Eq. (2.4) becomes

$$e^{\phi}(1 + r\phi') = 1$$

A change of variable from e^{ϕ} to γ, as described in Sokolnikoff's book [4], yields

$$\gamma + r\gamma' = 1$$

$$\frac{d(r\gamma)}{dr} = 1$$

$$r\gamma = r + \text{constant}$$

and

$$\gamma = 1 - \frac{2m}{r} = e^{\phi}$$

where $2m$ is a constant of integration. The metric tensor is thus

$$g_{ij} = \begin{bmatrix} -\left(1 - \frac{2m}{r}\right)^{-1} & 0 & 0 & 0 \\ 0 & -r^2 & 0 & 0 \\ 0 & 0 & -r^2 \sin^2 \theta & 0 \\ 0 & 0 & 0 & \left(1 - \frac{2m}{r}\right) \end{bmatrix}$$

The solution is complete once the constant m is determined. There are two ways of determining m. The first involves substituting the metric into the equation of geodesics and obtaining the equations of motion. The term for the weak gravity field containing m is equated with Newton's gravitational acceleration. This is a little tricky because the fourth coordinate of the geometry is ct and we need to factor out the c to get the equations of motion in terms of t. The result for the acceleration of r is given in Eq. (2.22) below and is

$$\frac{d^2r}{d\tau^2} = -\frac{mc^2}{r^2} + (r - 3m)\left(\frac{d\phi}{d\tau}\right)^2 \tag{2.7}$$

For a spacecraft being radially accelerated, the ϕ coordinate is constant and $d\phi$ is zero. Therefore,

$$\frac{d^2r}{d\tau^2} = -\frac{mc^2}{r^2} \tag{2.8}$$

The weak field acceleration according to Newton is

$$\frac{d^2r}{dt^2} = -\frac{GM}{r^2} = -\frac{\mu}{r^2}$$

In the weak field, $dt \approx d\tau$. Therefore, $mc^2 = \mu$ and

$$m = \frac{\mu}{c^2}$$

$$\frac{d^2r}{d\tau^2} = -\frac{\mu}{r^2} + \left(r - \frac{3\mu}{c^2}\right)\left(\frac{d\phi}{d\tau}\right)^2 \tag{2.9}$$

The same result may be obtained by solution of the Einstein field equations inside the sun. This will be made a lot easier by making the following substitutions in the Ricci tensor to temporarily get rid of the exponentials:

$$\Phi = e^\phi \quad \text{and} \quad \Lambda = e^\lambda$$

The Ricci tensor defined by Eqs. ((2.3)–(2.6)) becomes

$$R_{11} = \frac{\Phi''}{2\Phi} - \frac{\Phi'\Lambda'}{4\Phi\Lambda} - \frac{(\Phi')^2}{4\Phi^2} - \frac{1}{r}\frac{\Lambda'}{\Lambda}$$

$$R_{22} = \frac{r\Phi'}{2\Phi\Lambda} + \frac{1}{\Lambda} - \frac{r\Lambda'}{2\Lambda^2} - 1$$

$$R_{33} = \left(\frac{r\Phi'}{2\Phi\Lambda} + \frac{1}{\Lambda} - \frac{r\Lambda'}{2\Lambda^2} - 1\right)\sin^2\theta$$

$$R_{44} = -\frac{\Phi''}{2\Lambda} + \frac{\Phi'\Lambda'}{4\Lambda^2} + \frac{(\Phi')^2}{4\Phi\Lambda} - \frac{1}{r}\frac{\Phi'}{\Lambda}$$

The scalar curvature of space is given by

$$R = R^u_u = g^{uv} R_{uv} = g^{11} R_{11} + g^{22} R_{22} + g^{33} R_{33} + g^{44} R_{44}$$

$$R = -\frac{1}{\Lambda} R_{11} - \frac{1}{r^2} R_{22} - \frac{1}{r^2 \sin^2 \theta} R_{33} + \frac{1}{\Phi} R_{44}$$

Since in the Schwarzschild geometry the trajectory is planar, we may set $\theta = \frac{\pi}{2}$ and $R_{33} = R_{22}$.

$$R = -\frac{1}{\Lambda} R_{11} - \frac{2}{r^2} R_{22} + \frac{1}{\Phi} R_{44}$$

$$R = -\frac{\Phi''}{\Phi \Lambda} + \frac{\Phi' \Lambda'}{2 \Phi \Lambda^2} + \frac{(\Phi')^2}{2 \Phi^2 \Lambda} - \frac{2}{r} \frac{\Phi'}{\Phi \Lambda} + \frac{2}{r} \frac{\Lambda'}{\Lambda^2} + \frac{2}{r^2} \left(1 - \frac{1}{\Lambda} \right)$$

$$G_{44} = R_{44} - \frac{1}{2} g_{44} R \tag{2.10}$$

$$G_{44} = -\frac{\Phi''}{2\Lambda} + \frac{\Phi' \Lambda'}{4\Lambda^2} + \frac{(\Phi')^2}{4\Phi\Lambda} - \frac{1}{r} \frac{\Phi'}{\Lambda} + \frac{\Phi''}{2\Lambda} - \frac{\Phi' \Lambda'}{4\Lambda^2} - \frac{(\Phi')^2}{4\Phi\Lambda} + \frac{1}{r} \frac{\Phi'}{\Lambda}$$

$$+ \frac{1}{r} \frac{\Phi \Lambda'}{\Lambda^2} + \frac{\Phi}{r^2} \left(1 - \frac{1}{\Lambda} \right) \tag{2.11}$$

$$G_{44} = \left(\frac{1}{r^2} - \frac{e^{-\lambda}}{r^2} - \frac{1}{r} \frac{de^{-\lambda}}{dr} \right) e^\phi = 8\pi \frac{G}{c^2} T_{44} = 8\pi \frac{G}{c^2} \rho \tag{2.12}$$

Since

$$e^{-\lambda} = \left(1 - \frac{2m}{r} \right)$$

$$e^\phi = \left(1 - \frac{2m}{r} \right)$$

we get

$$2m = r - re^{-\lambda}$$

$$\frac{dm(r)}{dr} = \frac{1}{2} \left(1 - e^{-\lambda} - r \frac{de^{-\lambda}}{dr} \right)$$

After substituting into Eq. (2.12),

$$\frac{2\,dm}{r^2\,dr}\left(1 - \frac{dm}{r}\right) = 8\pi\frac{G}{c^2}\rho$$

$$dm\left(1 - \frac{dm}{r}\right) = 4\pi r^2\frac{G}{c^2}\rho\,dr$$

In the limit as dr and dm go to zero, the term in the bracket goes to one and can be factored out of the integral.

$$dm = 4\pi r^2\frac{G}{c^2}\rho\,dr$$

Starting at the center of the sun, we integrate over a thin spherical shell and

$$\Delta m = \frac{1}{c^2}\,\Delta\mu$$

The Δm is added to the m in G_{44}, and we continue adding spherical shells until we get to the surface (r_s). The differential line element associated with each spherical shell is given by

$$ds^2 = -\left\{\left(1 - \frac{2dm}{r}\right)^{-1}dr^2 + r^2\,d\theta^2 + r^2\sin^2\theta\,d\phi^2\right\} + \left(1 - \frac{2dm}{r}\right)c^2dt^2$$

(2.13)

and when we get to the surface, the constant $m(r_s)$

$$m = m(r_s) = \frac{Gm}{c^2} = \frac{\mu}{c^2} \tag{2.14}$$

can be inserted into the Schwarzschild–Ricci tensor.

The same result can be obtained by equating G_{11} with pressure in hydrostatic equilibrium.

$$G_{11} = \left(\frac{1}{r^2} - \frac{e^\phi}{r^2} - \frac{1}{r}\frac{de^\phi}{dr}\right)e^\lambda = 8\pi\frac{G}{c^2}T_{11} = 8\pi\frac{G}{c^2}p \tag{2.15}$$

$$\frac{dm(r)}{dr} = \frac{1}{2}\left(1 - e^\phi - r\frac{de^\phi}{dr}\right)$$

$$\frac{2\,dm}{r^2\,dr}\left(\frac{1}{1 - \frac{dm}{r}}\right) = 8\pi\frac{G}{c^2}p$$

Integrating over the volume gives the same result for $m(r_s)$ provided p is numerically equal to ρ. The Schwarzschild stress–energy tensor for a symmetrical non-rotating rigid sun is thus

$$T_{uv} = \begin{bmatrix} \rho & 0 & 0 & 0 \\ 0 & 0 & 0 & 0 \\ 0 & 0 & 0 & 0 \\ 0 & 0 & 0 & \rho \end{bmatrix}$$

The T_{11} term is equal to ρ because the sun is assumed to be rigid. The energy of compression and all other energy associated with motion are assumed to be included in ρ. The T_{22} and T_{33} terms are zero because there is no differential pressure normal to r. If a person stands up in a cave anywhere in the Earth, there is no sideways pressure causing him/her to fall over. The only pressure is in the r direction and it is felt by his/her feet. This pressure is equal to his weight or ρ *volume*. At the center of the Earth in a cave, a person experiences weightlessness provided the cave is strong enough to hold up the weight of the Earth. There is no singularity at the center of the Earth because the acceleration of gravity is zero.

Inside the sun the mass as a function of r is

$$m(r) = m(r_s)\frac{r^3}{r_s^3} \qquad (2.16)$$

The integration as a function of r stops at r and not at the surface. For spherical symmetry the mass located above r does not contribute to the acceleration. Therefore the space within a hollow spherical shell is Euclidean or flat space. Consider a hollow Earth or Sun where all the mass is concentrated in a thin shell. The geometry is illustrated in Fig. 2.1. In Newtonian mechanics the acceleration of a mass element is given by

$$a_y = \frac{\mu}{r^2}\sin\phi$$

where μ is the line density of a ring centered at y and parallel to the x–z plane. The acceleration of the ring is

$$a_y = \frac{\mu}{r^2}\sin\phi\, 2\pi r_s \cos\theta$$

Since

$$r^2 = (r_s \sin\theta - r_y)^2 + (r_s \cos\theta)^2$$

and

$$\sin\phi = \frac{r_s \sin\theta - r_y}{r}$$

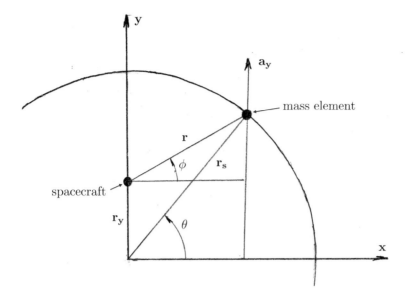

Fig. 2.1 Gravity of spherical shell

we have for a_y

$$a_y = \frac{2\pi r_s \cos\theta \, \mu (r_s \sin\theta - r_y)}{r^3}$$

The acceleration of the spacecraft due to the spherical shell or the acceleration of the spherical shell by the spacecraft is obtained by integrating a_y from θ equals -90 to 90 deg. The a_y component integrates to zero.

$$\bar{a}_y = \int_{\frac{-\pi}{2}}^{\frac{+\pi}{2}} a_y \, d\theta = \frac{2\pi \mu (r_s - r_y \sin\theta)}{r_y^2 (r_s^2 + r_y^2 - 2r_s r_y \sin\theta)^{\frac{1}{2}}} \Bigg|_{\frac{-\pi}{2}}^{\frac{\pi}{2}} = 0$$

This proves the shell theorem for $r < r_s$, which was first obtained by Newton. When r is greater than r_s, the acceleration is mass divided by r^2. The shell theorem is also true for general relativity. The proof is difficult because we cannot use the Euclidean model shown in Fig. 2.1. General relativity distorts the lengths of lines. However, we can adapt a geometrical proof given in the literature for Newtonian gravity. This is the proof of projected cones that is relatively easy to understand. The geometry is shown in Fig. 2.2. The acceleration in classical mechanics is the mass of the shell enclosed within the base of the cone divided by the distance to a point P squared. The acceleration of the point is thus independent of r because the r squared in the numerator cancels the r squared in the denominator. Thus the force of the cone defined by r_1 on P is equal and opposite to the force of the cone defined by r_2. The acceleration of P is thus zero when we extend the cone angle to 180 deg.

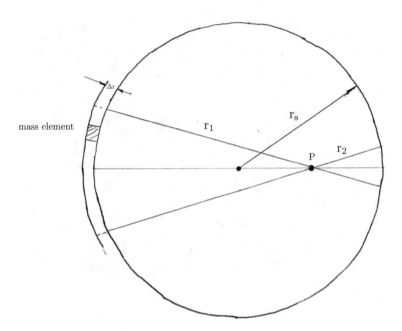

Fig. 2.2 General relativity gravity of spherical shell

For general relativity we use the same idea. However, the r coordinate is in curved space and the spherical angles are Euclidean. The time coordinate is also in curved space, but the sphere is static and there is no time variation. If we define a typical mass element, the mass in the base of the cones varies as the square of r. For this proof to work, the acceleration of the mass element must be inverse square. The acceleration of P caused by the mass element can be obtained from the exterior Schwarzschild solution Eq. (2.23).

$$\frac{d^2r}{d\tau^2} = -\frac{\mu}{r^2} + \left(r - \frac{3\mu}{c^2}\right)\left(\frac{d\phi}{d\tau}\right)^2 \tag{2.17}$$

Since the angular velocity is zero, the radial acceleration is inverse square. We move the point P to the center. Now the r_1 cone and the r_2 cone are equal and opposite due to symmetry. The acceleration is zero. Next we contract the shell until the radius approaches zero. Then we expand the shell a small δr. The acceleration of the point P within this shell is zero because of the shell theorem. We continue to expand the shell in small increments until we reach the surface. The acceleration of the mass elements within each incremental shell in the base of the r_1 cone is equal and opposite to the acceleration of the elements in the base of the r_2 cone. The elements in the r_1 cone are further away from the point P than the r_2 elements, but the r_1 elements are bigger than the r_2 elements. They cancel because of the inverse square relationship. The acceleration of P is zero for all points in the shell. Therefore, we can discard all the mass above r when integrating to define the metric tensor in a spherical body.

The metric tensor inside the sun can be obtained by inserting Eq. (2.16) for the mass distribution into Eq. (2.13) for the differential metric.

$$ds^2 = -\left\{ \left(1 - \frac{2m(r_s)r^3}{rr_s^3}\right)^{-1} dr^2 + r_s^2\, d\theta^2 + r_s^2 \sin^2\theta\, d\phi^2 \right\}$$
$$+ \left(1 - \frac{2m(r_s)r^3}{rr_s^3}\right) c^2 dt^2 \tag{2.18}$$

The radial acceleration inside the sun (Eq. (2.22)) is given by

$$\frac{d^2 r}{d\tau^2} = -\frac{m(r)c^2}{r^2} = \frac{\mu\, r}{r_s^3}$$

2.2 Schwarzschild Equations of Motion

With the metric defined, we are halfway there. The other half is developing equations of motion from the metric that may be programmed on a computer for navigation. The geodesic equation describes the acceleration of a particle in space–time coordinates and takes the place of the gradient in classical theory. Thus,

$$\frac{d^2 x^\alpha}{ds^2} + \Gamma^\alpha_{uv} \frac{dx^u}{ds}\frac{dx^v}{ds} = 0 \tag{2.19}$$

$$\Gamma^u_{\alpha\beta} = g^{uv}\Gamma_{v\alpha\beta}$$

$$\Gamma_{u\alpha\beta} = \frac{1}{2}\left(\frac{\partial g_{u\alpha}}{\partial x^\beta} + \frac{\partial g_{u\beta}}{\partial x^\alpha} - \frac{\partial g_{\alpha\beta}}{\partial x^u}\right)$$

The Christoffel symbols for the Schwarzschild solution are given by

$$\Gamma^r_{rr} = \Gamma^1_{11} = \frac{-m}{r^2}\left(1 - \frac{2m}{r}\right)^{-1}$$

$$\Gamma^r_{\theta\theta} = \Gamma^1_{22} = -r\left(1 - \frac{2m}{r}\right)$$

$$\Gamma^r_{\phi\phi} = \Gamma^1_{33} = -r\left(1 - \frac{2m}{r}\right)\sin^2\theta$$

$$\Gamma^r_{tt} = \Gamma^1_{44} = \frac{m}{r^2}\left(1 - \frac{2m}{r}\right)$$

$$\Gamma^t_{rt} = \Gamma^4_{14} = \frac{m}{r^2}\left(1 - \frac{2m}{r}\right)^{-1}$$

$$\Gamma^\phi_{r\phi} = \Gamma^3_{13} = \frac{1}{r}$$

$$\Gamma^\phi_{\theta\phi} = \Gamma^3_{23} = \cot\theta$$

$$\Gamma^\theta_{r\theta} = \Gamma^2_{12} = \frac{1}{r}$$

$$\Gamma^\theta_{\phi\phi} = \Gamma^2_{33} = -\sin\theta\cos\theta$$

To be consistent with the literature, the symbol for the Christoffel symbols $\{uv, \alpha\}$ has been changed to Γ^α_{uv}. It is no longer necessary to recognize that the Christoffel symbols are not a tensor. The equations of motion are obtained by substituting the Christoffel symbols into the geodesic equation. Since the motion is planar, we may rotate to a coordinate system such that the motion is in the x–y plane. The θ dependency is thus removed, and for $\theta = \frac{\pi}{2}$, we obtain from the geodesic equation

$$\frac{d^2r}{ds^2} = \frac{m}{r^2}\left(1 - \frac{2m}{r}\right)^{-1}\left(\frac{dr}{ds}\right)^2 + r\left(1 - \frac{2m}{r}\right)\left(\frac{d\phi}{ds}\right)^2$$

$$- \frac{m}{r^2}\left(1 - \frac{2m}{r}\right)\left(\frac{dct}{ds}\right)^2 \tag{2.20}$$

$$\frac{d^2\phi}{ds^2} = \frac{2}{r}\frac{dr}{ds}\frac{d\phi}{ds}$$

The acceleration defined by Eq. (2.19) is in curved space coordinates. The contravariant metric tensor (g^{uv}) that multiplies the Christoffel symbols transforms to curved space coordinates as shown in Eq. (1.28). The associated line element is Euclidean. In curved space the length of a second on the spacecraft is variable. We can define a time parameter (τ) where the second is constant in length as measured by a clock that is stationary and at infinity. If we assume a sun with a hollow core several thousand kilometers in diameter, we can place this clock at the sun's barycenter as shown in Fig. 2.2 since the hollow sphere has no gravity and is a flat Euclidean space. We can establish a relationship between coordinate time (t) and what is called affine parameter time (τ) from the line element defined by

$$ds^2 = -\left\{\left(1 - \frac{2m}{r}\right)^{-1} dr^2 + r^2\,d\theta^2 + r^2\sin^2\theta\,d\phi^2\right\} + \left(1 - \frac{2m}{r}\right)c^2dt^2$$

If we force the line element variation to be constant ($ds^2 = c^2\,d\tau^2$), the following equation relates a variation in t to a variation in τ.

$$\left(\frac{dt}{d\tau}\right)^2 = \left(1 - \frac{2m}{r}\right)^{-1} + \frac{1}{c^2}\left(1 - \frac{2m}{r}\right)^{-2}\left(\frac{dr}{d\tau}\right)^2 + \frac{r^2}{c^2}\left(1 - \frac{2m}{r}\right)^{-1}\left(\frac{d\phi}{d\tau}\right)^2$$

$$(2.21)$$

Substituting Eq. (2.21) into Eq. (2.20) gives the following equation of motion for a spacecraft:

$$\frac{d^2r}{d\tau^2} = -\frac{mc^2}{r^2} + (r - 3m)\left(\frac{d\phi}{d\tau}\right)^2 \tag{2.22}$$

Replacing m by $\frac{\mu}{c^2}$ as obtained from the solution of Einstein's tensor gives the following equations of motion:

$$\frac{d^2r}{d\tau^2} = -\frac{\mu}{r^2} + \left(r - \frac{3\mu}{c^2}\right)\left(\frac{d\phi}{d\tau}\right)^2 \tag{2.23}$$

$$\frac{d^2\phi}{d\tau^2} = -\frac{2}{r}\frac{dr}{d\tau}\frac{d\phi}{d\tau} \tag{2.24}$$

$$\left(\frac{dt}{d\tau}\right)^2 = \left(1 - \frac{2\mu}{c^2r}\right)^{-1} + \frac{1}{c^2}\left(1 - \frac{2\mu}{c^2r}\right)^{-2}\left(\frac{dr}{d\tau}\right)^2 + \frac{r^2}{c^2}\left(1 - \frac{2\mu}{c^2r}\right)^{-1}\left(\frac{d\phi}{d\tau}\right)^2$$

$$(2.25)$$

The trajectory of a photon differs from that of a particle or spacecraft moving at the speed of light even in the limit of very small mass for the spacecraft. The difference arises because a photon has zero rest mass, and thus there is no force of gravity acting on the photon that gives rise to Newtonian acceleration. The photon follows the contour of curved space. The resulting path is called the null geodesic.

We know from special relativity that an observer's clock on the photon will not register any passage of time. The proper time associated with a photon is simply the time that a stationary observer would measure. The difference of the affine parameter (τ) between two points times the speed of light is the distance that one would measure with a meter stick along the path of the photon. For a photon the measured length in curved space is zero. Since ds is zero, the geodesic equation degenerates to indeterminate forms that must be evaluated in the limit as ds goes to zero. The indeterminate form $ds^2/c^2d\tau^2$, which has the value of 1 for a spacecraft, has the value 0 for a photon in the limit as ds approaches zero.

$$\frac{ds^2}{d\tau^2} = 0 = -\frac{1}{c^2d\tau^2}\left\{\left(1 - \frac{2\mu}{c^2r}\right)^{-1}dr^2 + r^2\,d\theta^2 + r^2\sin^2\theta\,d\phi^2\right\}$$

$$+ \left(1 - \frac{2\mu}{c^2r}\right)\frac{dt^2}{d\tau^2}$$

The equations of motion for a photon are thus given by

$$\frac{d^2r}{d\tau^2} = \left(r - \frac{3\mu}{c^2}\right)\left(\frac{d\phi}{d\tau}\right)^2 \tag{2.26}$$

$$\frac{d^2\phi}{d\tau^2} = -\frac{2}{r}\frac{dr}{d\tau}\frac{d\phi}{d\tau} \tag{2.27}$$

$$\left(\frac{dt}{d\tau}\right)^2 = \frac{1}{c^2}\left(1 - \frac{2\mu}{c^2r}\right)^{-2}\left(\frac{dr}{d\tau}\right)^2 + \frac{r^2}{c^2}\left(1 - \frac{2\mu}{c^2r}\right)^{-1}\left(\frac{d\phi}{d\tau}\right)^2 \tag{2.28}$$

2.3 Isotropic Schwarzschild Coordinates

In the Newtonian world, before general relativity, the trajectories of the planets were observed through telescopes and the data fit to a model of the solar system based on Newton's equations of motion. From this model, the gravitational constant of the sun and the planetary ephemerides were estimated to an accuracy consistent with the measurement and model errors. With the introduction of general relativity to the model, the data was refit and a new set of constants and planetary ephemerides determined. However, since the relativistic effects are small, the differences between the numerical values associated with the curved space coordinates and the classical coordinates are also small. This small difference often results in confusion of the two coordinate systems.

In order to make the relativistic system more nearly coincide with the classical system, a coordinate transformation or a change of variable was devised to make the local curved space coordinates come into alignment with Euclidean coordinates. The volume element, which is a parallelepiped in curved space coordinates, is stretched and compressed to make it a cube. This transformation makes the relativistic coordinates look more classical but does not really change anything. The transformed coordinate system is called isotropic Schwarzschild coordinates. The transformation is given by

$$r = \left(1 + \frac{\mu}{2c^2\bar{r}}\right)^2 \bar{r} \tag{2.29}$$

$$\phi = \bar{\phi} \tag{2.30}$$

where $\bar{r}$ and $\bar{\phi}$ are the isotropic coordinates. In order to obtain the isotropic form of the equations of motion, we simply substitute the above equation for r into the exact Schwarzschild equations. The exact isotropic Schwarzschild line element is given by

$$ds^2 = \frac{\left(1 - \frac{\mu}{2c^2\bar{r}}\right)^2}{\left(1 + \frac{\mu}{2c^2\bar{r}}\right)^2} dt^2 - \frac{1}{c^2}\left(1 + \frac{\mu}{2c^2\bar{r}}\right)^4 \left(d\bar{r}^2 + r^2 d\bar{\phi}^2\right)$$

and this is approximated by

$$d\bar{s}^2 = \left(1 - \frac{2\mu}{c^2\bar{r}}\right) dt^2 - \frac{1}{c^2}\left(1 + \frac{2\mu}{c^2\bar{r}}\right)\left(d\bar{r}^2 + r^2 d\bar{\phi}^2\right)$$

The exact isotropic Schwarzschild equations of motion for a spacecraft become

$$\frac{d^2\bar{r}}{d\tau^2} = -\frac{\mu}{\bar{r}^2}\left(1 + \frac{\mu}{2c^2\bar{r}}\right)^{-4} + \left(1 - \frac{\mu^2}{4c^4\bar{r}^2}\right)^{-1}$$

$$\times \left\{\frac{\mu^3}{2c^4\bar{r}^5}\left(1 + \frac{\mu}{2c^2\bar{r}}\right)^{-4}\left(\frac{d\bar{r}}{d\tau}\right)^2 + \left[\left(1 + \frac{\mu}{2c^2\bar{r}}\right)^2\bar{r} - \frac{3\mu}{c^2}\right]\left(\frac{d\bar{\phi}}{d\tau}\right)^2\right\}$$

$$\frac{d^2\bar{\phi}}{d\tau^2} = -\frac{\left(1 - \frac{\mu^2}{4c^4\bar{r}^2}\right)}{\left(1 + \frac{\mu}{2c^2\bar{r}}\right)^2}\frac{2}{\bar{r}}\frac{d\bar{r}}{d\tau}\frac{d\bar{\phi}}{d\tau}$$

$$\left(\frac{d\bar{t}}{d\tau}\right)^2 = \frac{\left(1 + \frac{\mu}{2c^2\bar{r}}\right)^2}{\left(1 - \frac{\mu}{2c^2\bar{r}}\right)^2} + \frac{1}{c^2}\frac{\left(1 + \frac{\mu}{2c^2\bar{r}}\right)^6}{\left(1 - \frac{\mu}{2c^2\bar{r}}\right)^2}\left[\left(\frac{d\bar{r}}{d\tau}\right)^2 + \bar{r}^2\left(\frac{d\bar{\phi}}{d\tau}\right)^2\right]$$

and these may be approximated by

$$\frac{d^2\bar{r}}{d\tau^2} = -\frac{\mu}{\bar{r}^2}\left(1 - \frac{2\mu}{c^2\bar{r}}\right) + \left(\bar{r} - \frac{2\mu}{c^2}\right)\left(\frac{d\bar{\phi}}{d\tau}\right)^2 \tag{2.31}$$

$$\frac{d^2\bar{\phi}}{d\tau^2} = -\left(1 - \frac{\mu}{c^2\bar{r}}\right)\frac{2}{\bar{r}}\frac{d\bar{r}}{d\tau}\frac{d\bar{\phi}}{d\tau} \tag{2.32}$$

$$\left(\frac{d\bar{t}}{d\tau}\right)^2 = 1 + \frac{2\mu}{c^2\bar{r}} + \frac{1}{c^2}\left(1 + \frac{4\mu}{c^2\bar{r}}\right)\left[\left(\frac{d\bar{r}}{d\tau}\right)^2 + \bar{r}^2\left(\frac{d\bar{\phi}}{d\tau}\right)^2\right] \tag{2.33}$$

The exact isotropic Schwarzschild equations of motion for a photon become

$$\frac{d^2\bar{r}}{d\tau^2} = \left(1 - \frac{\mu^2}{4c^4\bar{r}^2}\right)^{-1}\left\{\frac{-\mu^2}{2c^4\bar{r}^3}\left(\frac{d\bar{r}}{d\tau}\right)^2 + \left[\left(1 + \frac{\mu}{2c^2\bar{r}}\right)^2\bar{r} - \frac{3\mu}{c^2}\right]\left(\frac{d\phi}{d\tau}\right)^2\right\}$$

$$\frac{d^2\bar{\phi}}{d\tau^2} = -\frac{\left(1 - \frac{\mu^2}{4c^4\bar{r}^2}\right)}{\left(1 + \frac{\mu}{2c^2\bar{r}}\right)^2}\frac{2}{\bar{r}}\frac{d\bar{r}}{d\tau}\frac{d\bar{\phi}}{d\tau}$$

$$\left(\frac{d\bar{t}}{d\tau}\right)^2 = \frac{1}{c^2}\frac{\left(1 + \frac{\mu}{2c^2\bar{r}}\right)^6}{\left(1 - \frac{\mu}{2c^2\bar{r}}\right)^2}\left[\left(\frac{d\bar{r}}{d\tau}\right)^2 + \bar{r}^2\left(\frac{d\bar{\phi}}{d\tau}\right)^2\right]$$

and these may be approximated by

$$\frac{d^2\bar{r}}{d\tau^2} = \left(\bar{r} - \frac{2\mu}{c^2}\right)\left(\frac{d\phi}{d\tau}\right)^2 \tag{2.34}$$

$$\frac{d^2\bar{\phi}}{d\tau^2} = -\left(1 - \frac{\mu}{c^2\bar{r}}\right)\frac{2}{\bar{r}}\frac{d\bar{r}}{d\tau}\frac{d\bar{\phi}}{d\tau} \tag{2.35}$$

$$\left(\frac{d\bar{t}}{d\tau}\right)^2 = \frac{1}{c^2}\left(1 + \frac{4\mu}{c^2\bar{r}}\right)\left[\left(\frac{d\bar{r}}{d\tau}\right)^2 + \bar{r}^2\left(\frac{d\bar{\phi}}{d\tau}\right)^2\right] \tag{2.36}$$

Chapter 3
Comparison of Numerical Integration and Analytic Solutions

A number of formulae are derived from the equations of motion that can be verified by experiment. They may also be verified by comparison with numerical integration. When the formulae are accurate enough they may be inserted into navigation software. Otherwise equations of motion must be programmed into the software and numerically integrated.

3.1 Mercury Perihelion Shift

Integration of the classical equations of motion for the orbit of Mercury reveals a shift in perihelion that cannot be accounted for with Newtonian theory. For navigation, it is necessary to modify the equations of motion to account for perihelion precession caused by the relativistic curvature of space near the sun. This is accomplished by use of a well-known formula or by numerical integration of the relativistic equations of motion. The results obtained by numerical integration can be compared with this formula. The well-known formula is found on the last page of Einstein's paper [1].

$$\delta\phi_0 = 24\pi^3 \frac{a^2}{T^2 c^2 (1 - e^2)}$$

and since the orbital period is

$$T = 2\pi \sqrt{\frac{a^3}{\mu_s}}$$

© The Author(s), under exclusive license to Springer Nature Switzerland AG 2021
J. Miller, C. J. Weeks, *General Relativity for Planetary Navigation*, SpringerBriefs
in Space Development, https://doi.org/10.1007/978-3-030-77546-9_3

we get the modern form of this equation.

$$\delta\phi_0 = \frac{6\pi\,\mu_s}{c^2 a(1 - e^2)}$$

where μ_s is the gravitational constant of the sun, a is the semi-major axis of Mercury's orbit, e is the orbital eccentricity, and c is the speed of light.

A simple derivation of the precession of Mercury's periapsis may be obtained by assuming that all the additional potential energy from general relativity goes into increasing the period of the orbit. The addition of the general relativity acceleration does not change the mean motion. After one revolution of the classical orbit, the perturbed orbit and the classical orbit have the same angular orientation because the orbits have the same angular momentum. At periapsis on the classical orbit, the perturbed orbit is descending for an additional δP to its periapsis. The precession is thus given by

$$\delta\phi_0 = 2\pi\,\frac{\delta P}{P}$$

$$\delta P = \frac{3P}{2a}\delta a$$

$$\delta a = \frac{a^2}{\mu}\,\delta C_3$$

and

$$\delta\phi_0 = \frac{2\pi}{P}\frac{3P}{2a}\frac{a^2}{\mu}\,\delta C_3 = \frac{3\pi a}{\mu}\,\delta C_3$$

From the Schwarzschild isotropic equations of motion (Eq. (2.31)), the radial acceleration is given by

$$\frac{d^2\bar{r}}{d\tau^2} = -\frac{\mu}{\bar{r}^2}\left(1 - \frac{2\mu}{c^2\bar{r}}\right)$$

Integrating the acceleration from $\bar{r}$ to infinity yields the potential energy and the General Relativity contribution is

$$\delta E_r = \frac{\mu^2}{c^2\bar{r}^2}$$

If the average radius ($\bar{r}^2$) is approximated by $b^2 = a^2(1 - e^2)$ the energy addition is

$$\delta C_3 = \frac{2\mu^2}{c^2 a^2(1 - e^2)} = 2\delta E_r$$

The factor of two is necessary because the energy orbit element (C_3) is twice the actual energy. Collecting terms, the Mercury precession is approximated by

$$\delta\phi_0 = \frac{3\pi a}{\mu} \frac{2\mu^2}{c^2 a^2 (1 - e^2)} = \frac{6\pi\mu}{c^2 a (1 - e^2)}$$

The equations of motion are integrated with the initial conditions computed from the state vector of Mercury at perihelion. After one complete revolution of Mercury about the sun, the integrated results are transformed to osculating orbit elements and the argument of perihelion is computed. In order to remove the integration error, the Newtonian equations of motion are integrated by the same numerical integrator in parallel with the relativistic equations of motion. The arguments of perihelion are differenced and compared with the formula. The same integration is repeated, only this time the isotropic form of the Schwarzschild equations of motion may be compared with the approximate formula. The results are displayed below.

Mercury Perihelion Shift

Perihelion shift formula	502.527×10^{-9} rad
Exact Schwarzschild integration	502.559×10^{-9} rad
Isotropic Schwarzschild integration	502.267×10^{-9} rad

The above results indicate that the formula for perihelion shift is quite accurate. The difference of 3×10^{-11} rad between the formula and the exact Schwarzschild integration may be attributed to the formula or perhaps integration error. The difference between the formula and the isotropic Schwarzschild integration is also small $(26 \times 10^{-11}$ rad$)$. This difference may also be attributed to integration error but may be the truncation error associated with the isotropic metric.

3.2 Radar Delay

The transit time of a photon or electromagnetic wave between two points in space is a measurement used to determine the orbits of the planets and spacecraft for the purposes of navigation and science. Both the navigation of a spacecraft and science experiments, particularly associated with General Relativity, require precise measurements of the transit time. Since the Deep Space tracking stations can measure times to within 0.1 ns, or about 3 cm, it is necessary to model the transit time to this accuracy.

The transit time of a photon or electromagnetic wave between two points in space is often referred to as the radar delay. This terminology originated with radar when a radio wave is transmitted and the delay in the reception of the reflected return is measured to determine the range. The time delay included that which is

associated with transmission media and the path length. Individual delay terms from
the troposphere, ionosphere, and solar plasma are identified and used to calibrate
the measured delay. For planetary spacecraft, the path length is computed from
the theory of General Relativity. For a round trip travel time, the additional delay
attributable to the curved space of General Relativity, over what would be computed
assuming flat space, can amount to approximately $250\,\mu s$.

$$ds^2 = \frac{\left(1 - \frac{\mu}{2c^2r}\right)^2}{\left(1 + \frac{\mu}{2c^2r}\right)^2} c^2 dt^2 - \left(1 + \frac{\mu}{2c^2r}\right)^4 \left(dr^2 + r^2 d\phi^2 + r^2 \sin^2\theta d\theta^2\right)$$

For a photon, $ds^2 = 0$ and the equation to be integrated for the elapsed coordinate
time (t) is obtained by transforming to Cartesian coordinates and solving for dt.

$$dt = \frac{1}{c} \frac{\left(1 + \frac{\mu}{2c^2r}\right)^3}{\left(1 - \frac{\mu}{2c^2r}\right)} \left(dx^2 + dy^2 + dz^2\right)^{\frac{1}{2}}$$

Expanding in a Taylor series and retaining terms of order c^{-5},

$$dt = \frac{1}{c}\left(1 + \frac{2\mu}{c^2r} + \frac{7}{4}\frac{\mu^2}{c^4r^2}\right)\left[dx^2 + dy^2 + dz^2\right]^{\frac{1}{2}} \tag{3.1}$$

The photon trajectory geometry is shown on Fig. 3.1. The motion is constrained
to the y–z plane and targeted from y_1, z_1 to y_2, z_2 such that the photon arrives at
the same y coordinate which is taken to be R. For this geometry, the x coordinate is
zero and the y coordinate variation is much smaller than the z coordinate variation.
Since for this problem $\frac{dy}{dz} \sim 10^{-4}$, the line element differentials are expanded as a
Taylor series,

$$\left(dx^2 + dy^2 + dz^2\right)^{\frac{1}{2}} \approx dz + \frac{1}{2}\frac{dy^2}{dz} + \mathcal{O}\left(\frac{dy^4}{dz^3}\right) \tag{3.2}$$

Changing the y variable of integration to z and inserting Eq. (3.2) into Eq. (3.1),

$$dt = \frac{1}{c}\left(1 + \frac{2\mu}{c^2r} + \frac{7}{4}\frac{\mu^2}{c^4r^2}\right)\left(dz + \frac{1}{2}\frac{dy^2}{dz^2}dz + \mathcal{O}\left(\frac{dy^4}{dz^3}\right)\right) \tag{3.3}$$

Fully expanded, there are nine terms in Eq. (3.3) and four of them are of order $1/c^5$
or greater. Consider a photon grazing the surface of the Sun. A maximum error of

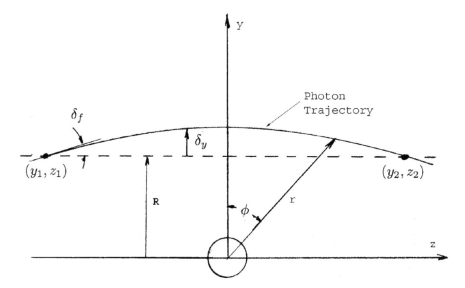

Fig. 3.1 Photon trajectory geometry

about 10 cm or 0.3 ns is desired. To achieve this accuracy, numerical integration of the equation of geodesics reveals that only four of the terms in Eq. (3.3) need be retained and these are,

$$
t_2 - t_1 = \frac{1}{c} \int_{z_1}^{z_2} \left[1 + \frac{2\mu}{c^2 r} + \frac{1}{2} \frac{dy^2}{dz^2} + \frac{7}{4} \frac{\mu^2}{c^4 r^2} \right] dz \tag{3.4}
$$

In carrying out the integration, care should be taken in geometrically interpreting the results. A "straight line" in curved space geometry, the shortest measured distance between two points, is the photon trajectory and not the dashed line shown on Fig. 3.1.

Consider the first term of Eq. (3.4),

$$
\Delta t_f = \frac{1}{c} \int_{z_1}^{z_2} dz = \frac{1}{c}(z_2 - z_1) \tag{3.5}
$$

This is called the flat space term. If the end points were in flat space, Δt_f would be the time a photon travels from point 1 on Fig. 3.1 to point 2. In curved space, there is no such thing as a straight line that connects these two points. The real interpretation of the term given by Eq. (3.5) is the mathematical result of performing the integration on the first term of Eq. (3.4).

The second term of Eq. (3.4) is called the logarithmic term for a reason that will become obvious.

$$\Delta t'_{log} = \frac{2\mu}{c^3} \int_{z_1}^{z_2} \frac{dz}{r}$$

Integration requires an equation for r as a function of z. An iterative solution may be obtained by assuming a solution for r and integrating to obtain a first approximation for t and y as a function of z. This solution is inserted into the remainder term, the difference between the assumed and actual function, and a second iterated solution may be obtained for t and y. This method of successive approximations can be continued until the required accuracy is achieved. As a starting function, "straight line" motion is assumed. Making use of the approximation that

$$r \approx \sqrt{z^2 + R^2}$$

$$\Delta t'_{log} = \frac{2\mu}{c^3} \int_{z_1}^{z_2} \left[\frac{1}{\sqrt{z^2 + R^2}} + \left(\frac{1}{r} - \frac{1}{\sqrt{z^2 + R^2}} \right) \right] dz = \Delta t_{log} + \Delta t_{rr} \qquad (3.6)$$

The first term of Eq. (3.6) integrates to the well-known equation for the time delay.

$$\Delta t_{log} = \frac{2\mu}{c^3} \ln \left[\frac{z_2 + \sqrt{z_2^2 + R^2}}{z_1 + \sqrt{z_1^2 + R^2}} \right] \qquad (3.7)$$

The second term of Eq. (3.6), which will be referred to as the radial remainder term (Δt_{rr}), requires a more accurate equation for the evaluation of r. In order to evaluate the terms associated with bending of the trajectory, an equation for y as a function of z is needed. The y coordinate is associated with the bending of the photon trajectory. Consider two photons in the plane of motion separated by ΔR. The plane containing these two photons and perpendicular to the velocity vector is the plane of the wave front. The bending is simply the distance one photon leads the other divided by their separation.

$$\delta = \frac{c \Delta t_d}{\Delta R}$$

In the limit as ΔR approaches zero, the equation for bending is

$$\delta = c \frac{dt_d}{dR}$$

The equation for the delay is taken to be the logarithmic term given by Eq. (3.7) and for simplicity the bending is computed starting at closest approach ($z_2 = 0$) to the origin.

$$t_d = \frac{2\mu}{c^3} \ln \left[\frac{z + \sqrt{z^2 + R^2}}{R} \right]$$

Taking the derivative with respect to R,

$$\frac{dt_d}{dR} \approx -\frac{2\mu}{c^2} \left\{ -\frac{1}{R} + \frac{R}{\sqrt{z^2 + R^2}(z + \sqrt{z^2 + R^2})} \right\}$$

Making use of the trigonometric approximations,

$$\cos\phi \approx \frac{R}{\sqrt{z^2 + R^2}}, \qquad \sin\phi \approx \frac{z}{\sqrt{z^2 + R^2}}$$

the equation for the bending reduces to

$$\delta = c \frac{dt_d}{dR} = -\frac{2\mu}{c^2 R} \frac{(\sin\phi + 1) - \cos^2\phi}{(\sin\phi + 1)} = -\frac{2\mu}{c^2 R} \sin\phi$$

$$\delta = \frac{2\mu}{c^2 R} \frac{z}{\sqrt{z^2 + R^2}}$$

Therefore, the accumulated bending from z_1 to z, expressed as differentials, is given by

$$\frac{dy}{dz} = \delta_f - \frac{2\mu}{c^2 R} \left(\frac{z}{\sqrt{z^2 + R^2}} - \frac{z_1}{\sqrt{z_1^2 + R^2}} \right) \tag{3.8}$$

where δ_f is the initial angle between the photon velocity vector and the horizontal line shown on Fig. 3.1. Referring to Fig. 3.1, the y component of the photon is

$$y = R + \delta_y$$

$$\delta_y(z) = \int_{z_1}^{z} \left(\delta_f - \frac{2\mu}{c^2 R} \left(\frac{z'}{\sqrt{z'^2 + R^2}} - \frac{z_1}{\sqrt{z_1^2 + R^2}} \right) \right) dz'$$

and

$$\delta_y = \delta_f(z - z_1) - \frac{2\mu}{c^2 R} \left(\sqrt{z^2 + R^2} - \frac{zz_1 + R^2}{\sqrt{z_1^2 + R^2}} \right) \tag{3.9}$$

The angle δ_f may be determined by evaluating the bending over the interval from z_1 to z_2. The coordinates are rotated to target the photon to the point $z = z_2$, where $\delta_y = 0$ and the constant gravitational aberration angle δ_f was determined as

$$\delta_f = \frac{1}{z_2 - z_1} \frac{2\mu}{c^2 R} \left(\sqrt{z_2^2 + R^2} - \frac{z_2 z_1 + R^2}{\sqrt{z_1^2 + R^2}} \right) \tag{3.10}$$

The angle δ_f simply rotates the coordinates of Fig. 3.1 such that y_1 and y_2 have the same value R.

The geometrical part of the radial remainder term, given in Eq. (3.6), may be approximated by making use of

$$\frac{1}{r} - \frac{1}{\sqrt{R^2 + z^2}} = \frac{1}{\sqrt{(R + \delta_y)^2 + z^2}} - \frac{1}{\sqrt{R^2 + z^2}} \approx \frac{-R\delta_y}{(R^2 + z^2)^{\frac{3}{2}}}$$

The complete radial remainder term (Δt_{rr}) is then given by

$$\Delta t_{rr} = -\frac{2\mu}{c^3} \int_{z_1}^{z_2} \frac{R}{(R^2 + z^2)^{\frac{3}{2}}} \left[\delta_f(z - z_1) - \frac{2\mu}{c^2 R} \left(\sqrt{z^2 + R^2} - \frac{z z_1 + R^2}{\sqrt{z_1^2 + R^2}} \right) \right] dz$$

$$\Delta t_{rr} = \frac{2\mu}{c^3 R^2} \left\{ \frac{1}{\sqrt{R^2 + z_2^2}} \left[\delta_f(R^3 + z_1 z_2 R) + \frac{2\mu R}{c^2} \arctan\left(\frac{z_2}{R}\right) \sqrt{R^2 + z_2^2} \right. \right.$$
$$\left. + \frac{2\mu}{c^2} \left(R^2 \sin\phi_1 + z_1 z_2 \sin\phi_1 - z_2 \sqrt{R^2 + z_1^2} \right) \right]$$
$$- \frac{1}{\sqrt{R^2 + z_1^2}} \left[\delta_f(R^3 + z_1^2 R) + \frac{2\mu R}{c^2} \arctan\left(\frac{z_1}{R}\right) \sqrt{R^2 + z_1^2} \right.$$
$$\left. \left. + \frac{2\mu}{c^2} \left(R^2 \sin\phi_1 + z_1^2 \sin\phi_1 - z_1 \sqrt{R^2 + z_1^2} \right) \right] \right\}$$

$$\Delta t_{rr} = \frac{2\mu}{c^3 R} \left\{ \delta_f \left[\frac{z_1 z_2 + R^2}{\sqrt{z_2^2 + R^2}} - \sqrt{z_1^2 + R^2} \right] \right.$$
$$\left. - \frac{2\mu}{c^2} \left[\arctan\left(\frac{z_1}{R}\right) - \arctan\left(\frac{z_2}{R}\right) + \frac{R(z_2 - z_1)}{\sqrt{z_1^2 + R^2}\sqrt{z_2^2 + R^2}} \right] \right\}$$

$$\tag{3.11}$$

The third term of Eq. (3.4) is the direct contribution of the trajectory bending to the time delay. This term is referred to as the bending term and is given by

$$\Delta t_b = \frac{1}{2c} \int_{z_1}^{z_2} \left(\frac{dy}{dz}\right)^2 dz$$

Substituting Eq. (3.8) for the slope into the above equation gives

$$\Delta t_b = \frac{1}{2c} \int_{z_1}^{z_2} \left(\delta_f + \frac{2\mu}{c^2 R}(\sin\phi - \sin\phi_1)\right)^2 dz$$

$$\Delta t_b = \frac{1}{2c} \int_{z_1}^{z_2} \left(\delta_f - \frac{2\mu}{c^2 R}\left(\frac{z}{\sqrt{z^2 + R^2}} - \frac{z_1}{\sqrt{z_1^2 + R^2}}\right)\right)^2 dz$$

Carrying out the integration

$$\Delta t_b = \frac{1}{2c^2 R}\left\{\left(\frac{4\mu^2}{c^4} + \delta_f^2 R^2 + \frac{4\mu\delta_f R}{c^2}\sin\phi_1 + \frac{4\mu}{c^4}\sin^2\phi_1\right)[z_2 - z_1]\right.$$

$$- \left(\frac{4\mu\delta_f R}{c^2} + \frac{8\mu^2}{c^4}\sin\phi_1\right)\left[\sqrt{R^2 + z_2^2} - \sqrt{R^2 + z_1^2}\right]$$

$$\left.+ \frac{4\mu^2 R}{c^4}\left[\arctan\left(\frac{R}{z_2}\right) - \arctan\left(\frac{R}{z_1}\right)\right]\right\}$$

$$\Delta t_b = \frac{1}{2c}\left\{\delta_f^2(z_2 - z_1) - \frac{4\mu}{c^2 R}\delta_f\left[\sqrt{z_2^2 + R^2} - \frac{z_1 z_2 + R^2}{\sqrt{z_1^2 + R^2}}\right]\right.$$

$$+ \frac{4\mu^2}{c^4 R^2}\left[\frac{R^2(z_1 + z_2) + 2z_1^2 z_2}{R^2 + z_1^2} - 2z_1\sqrt{\frac{z_2^2 + R^2}{z_1^2 + R^2}}\right.$$

$$\left.\left.+ R\left[\arctan\left(\frac{z_1}{R}\right) - \arctan\left(\frac{z_2}{R}\right)\right]\right]\right\} \tag{3.12}$$

The fourth and final term of Eq. (3.4) is the c^5 approximation to the error in the metric. This is a small term and contributes less than a nanosecond to the delay. The equation is given by

$$\Delta t_m = \frac{7}{4}\frac{\mu^2}{c^5}\int_{z_1}^{z_2}\frac{1}{r^2}dz \approx \frac{7}{4}\frac{\mu^2}{c^5}\int_{z_1}^{z_2}\frac{1}{R^2 + z^2}dz$$

Carrying out the integration

$$\Delta t_m \approx \frac{7}{4} \frac{\mu^2}{c^5 R} \left[\arctan\left(\frac{z_2}{R}\right) - \arctan\left(\frac{z_1}{R}\right) \right] \tag{3.13}$$

The complete equation for the coordinate time delay of a photon moving from (y_1, z_1) to (y_2, z_2) is obtained by summing all the individual terms and

$$t_2 - t_1 = \Delta t_f + \Delta t_{log} + \Delta t_{rr} + \Delta t_b + \Delta t_m \tag{3.14}$$

Before evaluating the individual terms of Eq. (3.14), the parameters used in the individual terms must be determined unambiguously from the end points of the photon trajectory. If two arbitrary end points in the y–z plane are defined by (y_1', z_1') and (y_2', z_2'), the vectors from the origin to these points are given by

$$\mathbf{r}_1 = (0, y_1', z_1') \qquad \text{and} \qquad \mathbf{r}_2 = (0, y_2', z_2')$$

and the vector from point 1 to point 2 is

$$\mathbf{r}_{12} = (0, y_2' - y_1', z_2' - z_1')$$

The angles between the vectors $\mathbf{r}_1$ and $\mathbf{r}_2$ and the vector $\mathbf{r}_{12}$ are computed from the dot products.

$$\phi_1 = \arccos\left(\frac{\mathbf{r}_1 \cdot \mathbf{r}_{12}}{r_2 r_{12}}\right), \qquad \phi_2 = \arccos\left(\frac{\mathbf{r}_2 \cdot \mathbf{r}_{12}}{r_2 r_{12}}\right)$$

The parameters needed in Eq. (3.14) with the coordinates rotated as shown on Fig. 3.1 are then given by

$$R = r_1 \sin\phi_1 = r_2 \sin\phi_2$$

$$z_1 = r_1 \cos\phi_1, \qquad z_2 = r_2 \cos\phi_2$$

and the angle δ_f is given by Eq. (3.6). The fully expanded equation for the transit time is given by,

$$t_2 - t_1 \approx \frac{1}{c}(z_2 - z_1) + \frac{2\mu}{c^3} \ln\left[\frac{z_2 + \sqrt{z_2^2 + R^2}}{z_1 + \sqrt{z_1^2 + R^2}}\right]$$

$$+ \frac{2\mu}{c^3 R}\left\{\delta_f \left[\frac{z_1 z_2 + R^2}{\sqrt{z_2^2 + R^2}} - \sqrt{z_1^2 + R^2}\right]\right.$$

$$
-\frac{2\mu}{c^2}\left[\arctan\left(\frac{z_1}{R}\right)-\arctan\left(\frac{z_2}{R}\right)+\frac{R(z_2-z_1)}{\sqrt{z_1^2+R^2}\sqrt{z_2^2+R^2}}\right]\Bigg\}
$$

$$
+\frac{1}{2c}\left\{\delta_f^2(z_2-z_1)-\frac{4\mu}{c^2R}\delta_f\left[\sqrt{z_2^2+R^2}-\frac{z_1z_2+R^2}{\sqrt{z_1^2+R^2}}\right]\right.
$$

$$
+\frac{4\mu^2}{c^4R^2}\left[\frac{R^2(z_1+z_2)+2z_1^2z_2}{R^2+z_1^2}-2z_1\sqrt{\frac{z_2^2+R^2}{z_1^2+R^2}}\right.
$$

$$
\left.\left.+R\left[\arctan\left(\frac{z_1}{R}\right)-\arctan\left(\frac{z_2}{R}\right)\right]\right]\right\}
$$

$$
+\frac{7}{4}\frac{\mu^2}{c^5R}\left[\arctan\left(\frac{z_2}{R}\right)-\arctan\left(\frac{z_1}{R}\right)\right]
$$

After simplification this equation takes the following form

$$
t_2-t_1\approx\frac{1}{c}(z_2-z_1)\left(1+\frac{1}{2}\delta_f^2\right)+\frac{2\mu}{c^3}\ln\left[\frac{z_2+\sqrt{z_2^2+R^2}}{z_1+\sqrt{z_1^2+R^2}}\right]
$$

$$
+\frac{2\mu}{c^3R}\delta_f\left[\frac{(z_1z_2-z_2^2)\sqrt{z_1^2+R^2}+(z_1z_2-z_1^2)\sqrt{z_2^2+R^2}}{\sqrt{z_1^2+R^2}\sqrt{z_2^2+R^2}}\right]
$$

$$
+\frac{2\mu^2}{c^5R^2}\left[\frac{R^2(z_1+z_2)+2z_1^2z_2}{z_1^2+R^2}-\frac{2z_2(z_1z_2+R^2)}{\sqrt{z_1^2+R^2}\sqrt{z_2^2+R^2}}\right]
$$

$$
+\frac{15}{4}\frac{\mu^2}{c^5R}\left[\arctan\left(\frac{z_2}{R}\right)-\arctan\left(\frac{z_1}{R}\right)\right]
$$

$$
\delta_f\approx\frac{1}{z_2-z_1}\frac{2\mu}{c^2R}\left(\sqrt{z_2^2+R^2}-\frac{z_2z_1+R^2}{\sqrt{z_1^2+R^2}}\right)\tag{3.15}
$$

Eq. (3.15) is the time delay associated with a photon or electromagnetic wave that passes through the gravitational field of a massive spherical body. The time delay is a function of only the gravitational constant of the massive body and the parameters z_1, z_2 and R, which may be computed directly from the isotropic Schwarzschild coordinates of the end points.

In order to determine the veracity of Eq. (3.15), a comparison with the time delay computed from numerical integration of the geodesic equations of motion was made and the result plotted on Fig. 3.2. In carrying out the numerical integration, a photon was initialized with a z coordinate of $-149,000,000$ km and y coordinate of

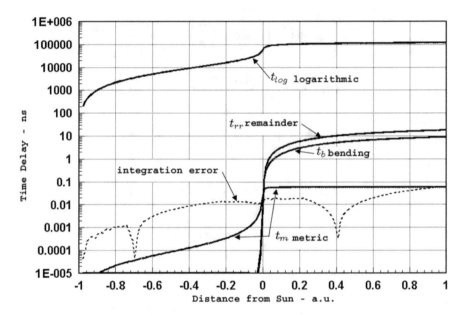

Fig. 3.2 Time delay for solar graze

696,000 km. The y component of velocity was set to zero and the z component to c. The x coordinates of position and velocity were set to zero. Thus the photon is initialized with a velocity magnitude equal to the speed of light and parallel to the z axis about 1 A.U. from the sun and on a flight path that would graze the surface of the sun if there were no bending due to General Relativity. The polar coordinates of the initial conditions were used to initialize the equations of motion and these were integrated by a fourth order Runge Kutta integrator with fifth order error control. The integration was stopped at various times along the flight path and Eq. (3.15) was evaluated. The required parameters were computed from the initial coordinates and the integrated coordinates at the time of the evaluation.

Also shown on Fig. 3.2 are some of the individual terms from Eq. (3.14). The linear term has been omitted since this term would require an additional 6 cycles of logarithmic scale. The dashed curve is the difference between the time delay computed by Eq. (3.15) and the results of numerical integration. This difference is attributed to error in the numerical integration algorithm. This was verified by setting the mass of the sun to zero and integrating straight line motion in the same coordinate system. Unfortunately, the integration error masked the error in the metric. Therefore, Eq. (3.15) could only be verified to about 0.05 ns which is about the same level of error as the error in the metric. The integration error of about $10^{-14} t$ is consistent with the error obtained integrating spacecraft orbits for navigation. Observe that the radial remainder term and bending term cause errors on the order of 10 ns or 37 cm.

3.3   Light Deflection

Light deflection is the bending of a photon or radio wave trajectory as it passes by a massive object. An experiment performed during a solar eclipse in 1919 measured the deflection of star light and was the first confirmation of General Relativity theory. For this comparison, we integrate the equations of motion for a photon and compare it with an analytic formula. The analytic formula is for a photon arriving at the Earth from infinity. This formula has been adapted to provide a continuous measure of the bending between any two points and is given by

$$\delta\phi = \frac{2\mu}{c^2 R}\{(\cos(90 + \phi_1) - \cos(90 + \phi_2)\}$$

where R is the closest approach to the sun, ϕ_1 is the angle from the y axis to the source and ϕ_2 is the angle from the y axis to the receiver. The y axis is in the direction of closest approach as illustrated on Fig. 3.1. Einstein's formula for the total bending is simply

$$\delta\phi_{12} = \frac{4\mu}{c^2 R}$$

where $\phi_1 = -90$ deg and $\phi_2 = 90$ deg.

Another formula for the bending is given by Eq. (3.10).

$$\delta_f = \frac{1}{z_2 - z_1}\frac{2\mu}{c^2 R}\left(\sqrt{z_2^2 + R^2} - \frac{z_2 z_1 + R^2}{\sqrt{z_1^2 + R^2}}\right)$$

If we take the limit as z_1 approaches minus infinity and z_2 approaches plus infinity, δ_f is one half of the Einstein bending formula. Since the total bending is the sum of the approach and departure bending, which are equal, the δ_f formula when multiplied by two is the Einstein formula. In order to get a formula for the total bending we have to apply Eq. (3.10) twice, once for the bending from z_1 to closest approach and once for the bending from closest approach to z_2. The total bending is thus

$$\delta_t = \frac{2\mu}{c^2 R}\left\{\frac{z_2}{\sqrt{z_2^2 + R^2}} - \frac{z_1}{\sqrt{z_1^2 + R^2}}\right\}$$

If z_1 is positive and equal to z_2, the bending is zero. If z_1 is negative and both z_1 and z_2 are much greater than R, we get Einstein's formula.

Another interesting application of the bending formula is the case where light is in a circular orbit about a concentrated mass or black hole. For a circular orbit, the local tangent is perpendicular to the radius. At closest approach z_1 is zero and

$$\delta_f = \frac{2\mu}{c^2 R} \left[\frac{z_2}{\sqrt{z_2^2 + R^2}} \right]$$

For a small z_2, the central angle ϕ shown on Fig. 3.1 is equal to δ_f. Angles are equal if they are perpendicular right side to right and left side to left. See your high school geometry book.

$$\Delta z_2 = R \Delta \delta_f = \frac{2\mu}{c^2} \left[\frac{\Delta z_2}{\sqrt{\Delta z_2^2 + R^2}} \right]$$

For a circular orbit, the radius is R. Thus we have

$$\Delta z_2 = \frac{2\mu}{c^2} \frac{\Delta z_2}{R}$$

$$R = \frac{2\mu}{c^2}$$

which makes R the event horizon of a black hole.

Comparison of the Einstein formula with numerical integration of the isotropic Schwarzschild equations of motion is a little tricky because it is necessary to define what is meant by bending in curved space. The generally accepted definition is the angle between the local tangent of the photon trajectory and the straight line path that the photon would follow if the sun was removed. Thus, in isotropic Schwarzschild coordinates, the deflection is given by

$$\delta\phi = \tan^{-1} \left(\frac{V_r \cos\phi - V_n \sin\phi}{V_r \cos\phi + V_n \sin\phi} \right)$$

where

$$V_r = \frac{dr}{d\tau}$$

$$V_n = r\frac{d\phi}{d\tau}$$

and the undeflected photon is assumed to move parallel to the z axis.

The equations of motion are initialized with the position and velocity of the photon. We place the photon far from the sun on a trajectory that will graze the surface of the sun. The initial state vector is given by

$$r_1 = 149,001,625.\,\text{km}$$

$$\phi_1 = -89.73236 \deg$$

$$\frac{dr_1}{d\tau} = -299{,}789.729 \, \text{km/s}$$

$$\frac{d\phi_1}{d\tau} = 9.3982872536 \times 10^{-6} \, \text{rad/s}$$

and the constants are

$$\mu = 1.327124399 \times 10^{11} \, \text{km}^3/\text{s}^2$$

$$c = 299{,}792.458 \, \text{km/s}$$

The equations of motion are integrated along a trajectory that grazes the sun and terminates at

$$\tau_2 = 954.901039554 \, \text{s (affine parameter time)}$$

$$t_2 = 954.901158130 \, \text{s (coordinate time)}$$

$$r_2 = 137{,}274{,}407 \, \text{km}$$

$$\phi_2 = 89.70998749 \, \text{deg}$$

$$\frac{dr_2}{d\tau} = 299{,}789.146 \, \text{km/s}$$

$$\frac{d\phi_2}{d\tau} = 11.072650234 \times 10^{-6} \, \text{rad/s}$$

A comparison of the total bending obtained by numerical integration with the theoretical formula derived by Einstein gives

Total Light Deflection Angle

Einstein's formula	8.48622×10^{-6} rad
Exact Schwarzschild integration	8.48642×10^{-6} rad

3.4   Clock Time Keeping

According to the theory of special theory, a clock running in a frame of reference that is moving with respect to an observers frame of reference will appear to run slower to the observer. According to the theory of general relativity, a clock running in a gravitational potential field will run slower than a clock outside the field.

Therefore, an observer that is stationary with respect to the solar system, will see the atomic clocks at the tracking stations running slower than his hypothetical clock. The observer is placed stationary with respect to the barycenter of the solar system because the equations of motion are written with respect to this center and placed far away to escape the effect of the gravitational acceleration of the Sun and planets. The coordinate time thus defined is called post-Newtonian time (PNT).

The relationship between PNT and the proper time measured by an atomic clock is given by the metric. For a particle moving in an orbit around the Sun, the metric in isotopic Schwarzschild coordinates is given by,

$$ds^2 = \frac{\left(1 - \dfrac{U}{2c^2}\right)^2}{\left(1 + \dfrac{U}{2c^2}\right)^2} c^2 dt^2 - \left(1 + \frac{U}{2c^2}\right)^4 \left(dx^2 + dy^2 + dz^2\right)$$

Retaining terms to order c^2, the metric may be approximated by,

$$ds^2 = \left(1 - \frac{2U}{c^2}\right) c^2 dt^2 - v^2 \, dt^2$$

where

$$v^2 = \left(\frac{dx}{dt}\right)^2 + \left(\frac{dy}{dt}\right)^2 + \left(\frac{dz}{dt}\right)^2$$

Solving for proper time ($ds^2 = c^2 \, d\tau^2$) we obtain

$$\frac{d\tau}{dt} = \sqrt{1 - \frac{2U}{c^2} - \left(\frac{v}{c}\right)^2}$$

which may be further approximated by

$$\frac{d\tau}{dt} = 1 - \frac{\mu_s}{c^2 r} - \frac{1}{2}\frac{v^2}{c^2} - \frac{\mu_e}{c^2 r_e}$$

where the Earth's gravitational potential is separated from the Sun's. The atomic clock time (τ) is obtained as a function of t by integrating the metric in conjunction with the equations of motion.

$$\tau = \int_{t_0}^{t} (1 - L) \, dt$$

where

$$L = \frac{\mu}{c^2 r} + \frac{1}{2} \frac{v^2}{c^2} + \frac{\mu_e}{c^2 r_e}$$

The function L can be separated into a constant term (L_0), secular terms that grow with time (L_s), and periodic terms (L_p). Thus we have,

$$L = L_0 + L_s + L_p$$

The constant term (L_0) is obtained by averaging L over all time and can be represented by,

$$L_0 = \frac{1}{c^2} \left(\frac{\mu}{r_0} + \frac{1}{2} v_0^2 \right) + \frac{\mu_e}{c^2 r_e}$$

where r_0 and v_0 are constants that give the correct average value for L_0. For the Earth's orbit about the Sun, r_0 is approximately the semi-major axis of the orbit and v_0 approximately the mean orbital velocity. Since the orbit is nearly an ellipse,

$$\frac{\mu}{a} = \frac{2\mu}{r} - v^2$$

and for $r = a$,

$$L_0 \approx \frac{3\mu}{2c^2 a} + \frac{\mu_e}{c^2 r_e}$$

The secular terms L_s are assumed to be zero because of conservation of energy and momentum. This leaves the periodic terms and these are given by

$$L_p = \frac{1}{c^2} \left(\frac{\mu}{r} - \frac{\mu}{r_0} + \frac{1}{2} v^2 - \frac{1}{2} v_0^2 \right)$$

and

$$\tau = t + \int_{t_0}^{t} -L_0 - \frac{1}{c^2} \left(\frac{\mu}{r} - \frac{\mu}{r_0} + \frac{1}{2} v^2 - \frac{1}{2} v_0^2 \right) dt \tag{3.16}$$

An approximate analytic formula for the periodic terms, derived by Brooks Thomas, is given by

$$\tau \approx t - L_0 (t - t_0) - \frac{2}{c^2} (\dot{\mathbf{r}}_b^s \cdot \mathbf{r}_b^s) - \frac{1}{c^2} (\dot{\mathbf{r}}_b^c \cdot \mathbf{r}_e^b) - \frac{1}{c^2} (\dot{\mathbf{r}}_e^c \cdot \mathbf{r}_{sta}^e) - \frac{1}{c^2} (\dot{\mathbf{r}}_s^c \cdot \mathbf{r}_b^s)$$

$$- \frac{\mu_j}{c^2 (\mu_j + \mu_s)} (\dot{\mathbf{r}}_j^s \cdot \mathbf{r}_j^s) - \frac{\mu_{sa}}{c^2 (\mu_{sa} + \mu_s)} (\dot{\mathbf{r}}_{sa}^s \cdot \mathbf{r}_{sa}^s) \tag{3.17}$$

In the notation used above, the position of the body identified by the subscript is with respect to the body identified by the superscript, where $c = $ the solar system barycenter, $s = $ the Sun, $b = $ the Earth-Moon barycenter, $e = $ the Earth, sta $= $ station location, $j = $ Jupiter, and $sa = $ Saturn.

The constant term (L_0) is the major contributor to the difference between ephemeris time and solar system barycenter time. If all the planets were in circular orbits and the tracking station was located at the Earth barycenter we would be done. Since the Earth-Moon barycenter is inside the Earth it would be difficult to track the spacecraft. The periodic terms account for the eccentricity of the planet orbits, the rotation of Earth on its axis and station locations.

3.5 Solar Pressure

Photons emanating from the sun impinge on the spacecraft resulting in a force that accelerates the spacecraft. The force results from the change in momentum as the photon decelerates from the speed of light (c) to rest with respect to the spacecraft and is converted to heat assuming the spacecraft is a black body. The force related to the change in linear momentum is given by,

$$F = \frac{d(mc)}{dt} \tag{3.18}$$

The incremental energy required to decelerate the photon is

$$dE = F dx$$

and the power supplied to the spacecraft is obtained by dividing by dt.

$$\frac{dE}{dt} = Fc \tag{3.19}$$

Over the time interval dt, the photon moves an infinitesimal amount dx. The mathematics are a bit over simplified but, if Eq. (3.19) is substituted into Eq. (3.18) we get $E = mc^2$ and this result is consistent with Special Relativity. For Newtonian mechanics

$$\frac{dE}{dt} = \frac{d(mv)}{dt} \frac{dx}{dt}$$

$$dE = d(mv) \frac{dx}{dt} = v \, d(mv)$$

$$E = m \int_0^c v dv = \frac{1}{2} mc^2$$

The classical solar pressure is thus one half of the General Relativity result. Here, Special Relativity is included in General Relativity. Since the power absorbed by a spacecraft, the mass of the spacecraft, the velocity of light, and the acceleration of a spacecraft can be determined with high precision, experiment verifies General Relativity. Since the force of a photon on the spacecraft and the interval of time and length traveled during contact are not needed, we are left with an empty feeling that this result could have been predicted by some other theory. The difference is only a factor of two.

Since $E = mc^2$ is such an important equation, it may be interesting to derive this result from the relativistic force a photon exerts on a body. Since a spacecraft is accelerated, a photon must exert a force over some distance. In Reference [4], Sokolnikoff provided an excellent derivation of $E = mc^2$ on pages 292 through 297 starting with the metric tensor. The mathematics can be found in many textbooks and are probably the property of Einstein. Most of the explanations are not as clear as Sokolnikoff's. However, the mathematics are a bit tedious.

The Minkowski metric can be obtained from Schwarzschild's metric by transforming to Cartesian coordinates and assuming zero mass.

$$ds^2 = c^2 dt^2 - (dx^2 + dy^2 + dz^2)$$

$$ds = \sqrt{c^2 - v^2}\, dt$$

The observed velocity associated with ds is

$$u^\alpha = \frac{dx^\alpha}{ds} = \frac{1}{\sqrt{c^2 - v^2}} \frac{dx^\alpha}{dt}$$

The relativistic force in curved space coordinates is defined by Newton's second law.

$$F^\alpha = \frac{\delta}{\delta s}(m_0 u^\alpha)$$

$$F^\alpha = \frac{1}{\sqrt{c^2 - v^2}} \frac{\delta}{\delta t}\left(m_0 \frac{dx^\alpha}{ds}\right)$$

$$F^\alpha = \frac{1}{\sqrt{c^2 - v^2}} \frac{\delta}{\delta t}\left(\frac{m_0}{\sqrt{c^2 - v^2}} \frac{dx^\alpha}{dt}\right)$$

$$F^\alpha = \frac{1}{c^2\sqrt{1 - \frac{v^2}{c^2}}} \frac{\delta}{\delta t}\left(\frac{m_0}{\sqrt{1 - \frac{v^2}{c^2}}} \frac{dx^\alpha}{dt}\right)$$

In order to keep Newton's second law valid, mass is defined by

$$m = \frac{m_0}{\sqrt{1 - \frac{v^2}{c^2}}}$$

$$F = c^2 \sqrt{1 - \frac{v^2}{c^2}} F^\alpha = \frac{\delta}{\delta t}\left(m \frac{dx^\alpha}{dt}\right) = ma^\alpha$$

The kinetic energy, which is conserved, is obtained by integrating the force over the distance traveled by the photon during contact.

$$E_k = \int_{P_0}^{P} F_i \, dx^i = \int_{P_0}^{P} \frac{\delta}{\delta t}\left(\frac{m_0 v^i}{\sqrt{1 - \frac{v^2}{c^2}}}\right) dx^i$$

A change of variable from P_0 to t yields

$$E_k = m_0 \int_{t_0}^{t} \left[\frac{d}{dt}\left(\frac{1}{\sqrt{1 - \frac{v^2}{c^2}}}\right) v^i \frac{dx^i}{dt} + \frac{dv^i}{dt}\left(\frac{1}{\sqrt{1 - \frac{v^2}{c^2}}}\right)\frac{dx^i}{dt}\right] dt$$

If we define

$$\beta^2 = \frac{v^2}{c^2} \quad v^i = \frac{dx^i}{dt}$$

$$\beta^2 c^2 = v_i \frac{dx_i}{dt} \quad \beta\dot\beta = \frac{v_i}{c^2}\frac{dv_i}{dt}$$

we get

$$E_k = m_0 \int_{t_0}^{t} \left[\frac{d}{dt}\left(\frac{1}{\sqrt{1 - \beta^2}}\right)\beta^2 c^2 + \left(\frac{c^2\beta\dot\beta}{\sqrt{1 - \beta^2}}\right)\right] dt$$

$$E_k = m_0 \int_{t_0}^{t} \left[\left(\frac{\beta\dot\beta}{(1 - \beta^2)^{\frac{3}{2}}}\right)\beta^2 c^2 + \left(\frac{c^2\beta\dot\beta}{(1 - \beta^2)^{\frac{1}{2}}}\right)\right] dt$$

$$E_k = m_0 c^2 \int_{t_0}^{t} \frac{\beta\dot\beta}{(1 - \beta^2)^{\frac{3}{2}}} dt$$

Since $\dot\beta = \frac{d\beta}{dt}$

$$E_k = m_0 c^2 \int_{P_0}^{P} \frac{\beta d\beta}{(1-\beta^2)^{\frac{3}{2}}}$$

$$E_k = m_0 c^2 \int_{P_0}^{P} d \left[\frac{1}{(1-\beta^2)^{\frac{1}{2}}} \right]$$

and

$$E_k = \frac{m_0 c^2}{(1-\frac{v^2}{c^2})^{\frac{1}{2}}} + \text{constant}$$

The kinetic energy is thus

$$E_k = mc^2 - m_0 c^2$$

and the total energy is

$$E = mc^2$$

For a photon, the rest mass (m_0) is zero. For a particle with mass, the rest energy is the intrinsic energy required to form the particle during the big bang.

The force of photons on the spacecraft creates a pressure over the exposed surface area (A) and the net force is obtained by integrating all the photons over the projected area (A) of the spacecraft.

$$F = \frac{1}{c} \frac{dE}{dt}$$

$$\frac{dE}{dt} = I A$$

The power supplied to the spacecraft per unit area (I) may be computed from the solar intensity measured at Earth (I_e) and scaled by the inverse square of the distance from the sun.

$$I = \left(\frac{R_e}{R_s} \right)^2 I_e \tag{3.20}$$

Collecting terms and solving for the force on the spacecraft gives

$$F = \frac{K A}{R_s^2} \tag{3.21}$$

Table 3.1 Solar pressure model parameters

	Front side	Cylindrical side	Back side	Antenna
Area	$8.92\,\text{m}^2$	$2.22\,\text{m}^2$	$11.25\,\text{m}^2$	$2.33\,\text{m}^2$
γ	0.165	0.039	0.750	0.750
β	0.742	0.101	0.100	0.107
ϵ	-0.112	0.400	0.400	0.398

$$K = \frac{1}{c} R_e^2 I_e$$

where

$$I_e = 1,353\ \text{w/m}^2,\ c = 2,999,793.458\ \text{m/s}^2 \text{ and } R_e = 149.4 \times 10^9\ \text{m}$$

and

$$K = 1.01 \times 10^{17}\ \text{kg m/s}^2 = 1.01 \times 10^8\ \text{kg km}^3/\text{m}^2\text{s}^2$$

K is given in both MKS units and mixed units since area is generally given in m^2 and distance in km. The solar pressure model used for navigation is more complicated than the simple flat plate black body model described above suggests. The above result for the force assumes that all the photons are absorbed by the spacecraft and are directed radially away from the Sun. Some of the photons will be reflected from the spacecraft which will increase the solar pressure. If the spacecraft were a perfect mirror the force would be doubled. The incident and reflected momentum exchange would be in the same direction. The reflected solar energy is composed of specular and diffuse radiation. For specular reflection the angle of incidence is equal to the angle of reflection and for diffuse radiation the energy is scattered by the cosine of the sun angle. The solar pressure model used for the Near Earth Asteroid Rendezvous (NEAR) mission had three components. The specular radiation component is γ, the diffuse radiation component is β and the third component ϵ accounts for thermal reradiation. Table 3.1 gives the values of these coefficients for each part of the spacecraft. A separate set of coefficients was specified for the front side, cylindrical side, back side and antenna.

3.6 Planetary and Stellar Aberration

The observed direction of light from a distance source differs from the actual direction obtained by solution of the light time equation due to the velocity parallax of the observer with respect to the photons or incoming wave front. This velocity parallax is referred to as aberration by astronomers and is aptly named. The first definition of aberration in the Webster's second edition dictionary is deviation from what is right, natural, or normal. Light is red shifted or blue shifted in frequency

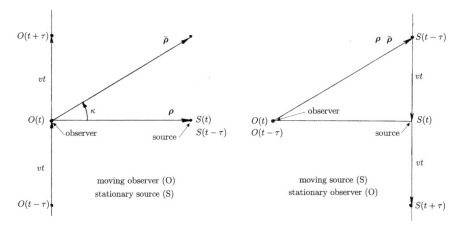

Fig. 3.3 Observer and source relative motion

depending whether the source is moving away from or toward the observer. The Doppler frequency shift is determined by the relative velocity. The observed direction of the light wave front is also affected by the motion of the source with respect to the observer but only the source velocity contributes to aberration. This apparent contradiction of Special Relativity may be resolved by examining the light time solution in conjunction with aberration in an inertial frame. Consider the case of an observer moving with respect to a stationary source as shown on the left side of Fig. 3.3. Assume that closest approach occurs at time t. The observer at time t, identified by $O(t)$, will receive an incoming light wave from the direction ρ which is the solution of the light time equation. The observer was at $O(t - \tau)$ when the photons were emitted by the source so aberration should not be confused with the solution of the light time equation. Because of aberration, due to the relative motion of the source with respect to the photons, the observer will see the source in the direction defined by the vector $\bar{\rho}$. The geometry is analogous to rain drops falling straight down. If the person starts to move, the rain drops appear slanted with respect to the local vertical.

Now suppose the observer is stationary and the relative motion of the source with respect to the observer is the same. In this case, shown on the right side of Fig. 3.3, the source appears to be moving in the opposite direction with velocity v. The observer at time t sees a plane wave emanating from the location of the source at $t - \tau$. There is no aberration. This is consistent with Special Relativity because the observer sees the source in the same direction for both cases. Since the Doppler shift is dependent on the relative velocity, the observer sees the same Doppler shift for both sides of Fig. 3.3. Furthermore, the speed of light will be the same for both observations due to time dilation and Lorentz contraction.

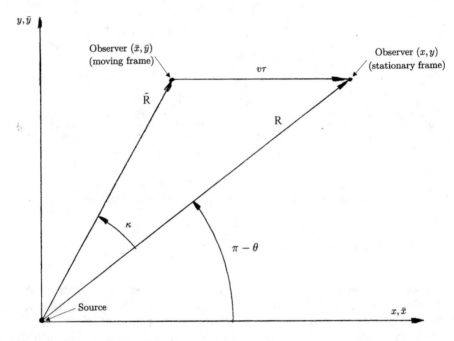

Fig. 3.4 Observer velocity relative to source

The geometry associated with an observer moving with respect to a source is illustrated on Fig. 3.4. In the stationary inertial frame, the source emits a photon at time $t = 0$ from the origin and this photon arrives at the observer at time $t = \tau$ with coordinates (x, y). The light time solution vector is $\mathbf{R}$ and the travel time is given by

$$\tau = \frac{R}{c}$$

If a coordinate system is defined that is moving with the observer $(\bar{x}, \bar{y})$ in the $+x$ direction with velocity v and the origins coincide at time $t = 0$, the observed direction of the photon is given by $\bar{\mathbf{R}}$. The angle between $\mathbf{R}$ and $\bar{\mathbf{R}}$ is the aberration angle κ and

$$\sin(\kappa) = \frac{|\mathbf{R} \times \bar{\mathbf{R}}/|}{R\,\bar{R}} \tag{3.22}$$

The angle θ is between the velocity vector and the vector from the observer to the source $(-\mathbf{R})$.

If classical Galilean motion is assumed, where the speed of light is not constant, the observed vector is given by

$$\bar{\mathbf{P}} = \begin{bmatrix} 1 & 0 & 0 & 0 \\ -v & 1 & 0 & 0 \\ 0 & 0 & 1 & 0 \\ 0 & 0 & 0 & 1 \end{bmatrix} \quad \mathbf{P} = \begin{bmatrix} \dfrac{R}{c} \\ -\dfrac{Rv}{c} - R\cos\theta \\ R\sin\theta \\ 0 \end{bmatrix} \tag{3.23}$$

where time is artificially carried along as the first component of $\mathbf{P}$ and $\mathbf{R}$ is contained in the last three components of $\mathbf{P}$.

$$\mathbf{P} = \begin{bmatrix} \dfrac{R}{c} \\ -R\cos\theta \\ R\sin\theta \\ 0 \end{bmatrix}$$

The angle κ is obtained by substituting the position vectors obtained from the second through third components of $\mathbf{P}$ and $\bar{\mathbf{P}}$ into Eq. (3.22).

$$\sin\kappa = \frac{v\sin\theta}{c} \left\{ \frac{1}{\sqrt{1 + \dfrac{v^2}{c^2} + \dfrac{2v\cos\theta}{c}}} \right\}$$

Making use of the approximation

$$\frac{1}{\sqrt{1 + \dfrac{v^2}{c^2} + \dfrac{2v\cos\theta}{c}}} \approx 1 - \frac{v\cos\theta}{c} - \frac{v^2}{2c^2}$$

and

$$\sin\theta\cos\theta = \frac{1}{2}\sin(2\theta)$$

the aberration angle may be approximated to second order by

$$\sin\kappa \approx \frac{v}{c}\sin\theta - \frac{1}{2}\frac{v^2}{c^2}\sin(2\theta) + \cdots$$

The classical result assumes that the speed of light in the moving frame is different from the speed of light in the inertial frame at rest. The Lorentz transformation from Special Relativity is used to get the correct result.

$$
\bar{P} =
\begin{bmatrix}
\dfrac{1}{\sqrt{1 - \dfrac{v^2}{c^2}}} & \dfrac{-v}{c^2\sqrt{1 - \dfrac{v^2}{c^2}}} & 0 & 0 \\[2em]
\dfrac{-v}{\sqrt{1 - \dfrac{v^2}{c^2}}} & \dfrac{1}{\sqrt{1 - \dfrac{v^2}{c^2}}} & 0 & 0 \\[2em]
0 & 0 & 1 & 0 \\[0.5em]
0 & 0 & 0 & 1
\end{bmatrix} P
$$

The Lorentz transformation is given in conventional engineering coordinates where time has the units of time and the existence of c is explicitly acknowledged. Since c is a constant, a system of space-time coordinates can be defined with $c = 1$ and time given the dimension of length. For these coordinates, favored by relativists, the Lorentz transformation matrix is symmetrical. The advantage of the conventional coordinates used here is that it is immediately obvious that the approximation to first order reduces to the Galilean transformation given by Eq. (3.23) in the limit as c approaches infinity. Since the "at rest" coordinate system is arbitrary, the inverse of the Lorentz transformation matrix can be obtained by changing the sign of v. The Galilean transformation also has this property. Another property of the Lorentz transformation is that the Minkowski metric must be preserved.

$$
ds^2 = c^2 dt^2 - dx^2 - dy^2 - dz^2 \tag{3.24}
$$

Since ds^2 is null for a photon ($ds^2 = 0$) then $d\bar{s}^2$ must also be null. The observation vector in the moving frame is given by

$$
\bar{P} =
\begin{bmatrix}
\dfrac{R}{c\sqrt{1 - \dfrac{v^2}{c^2}}} + \dfrac{Rv\cos\theta}{c^2\sqrt{1 - \dfrac{v^2}{c^2}}} \\[2em]
-\dfrac{Rv}{c\sqrt{1 - \dfrac{v^2}{c^2}}} - \dfrac{R\cos\theta}{\sqrt{1 - \dfrac{v^2}{c^2}}} \\[2em]
R\sin\theta \\[0.5em]
0
\end{bmatrix}
$$

When $\bar{P}$ is substituted into the Minkowski metric (Eq. (3.24)) it is demonstrated that $d\bar{s}^2$ is null. Proceeding as for the classical solution, the angle κ corrected for Special Relativity is obtained.

$$
\kappa = \arcsin\left(\frac{\dfrac{v\sin\theta}{c} + \left(1 - \sqrt{1 - \dfrac{v^2}{c^2}}\right)\sin\theta\cos\theta}{1 + \dfrac{v}{c}\cos\theta} \right)
$$

Making use of the approximations

$$\sqrt{1 - \frac{v^2}{c^2}} \approx 1 - \frac{1}{2}\frac{v^2}{c^2}$$

$$\frac{1}{1 + \frac{v}{c}\cos\theta} \approx 1 - \frac{v}{c}\cos\theta$$

the first two terms of the series expansion for $\sin\kappa$ are

$$\sin\kappa \approx \frac{v}{c}\sin\theta - \frac{1}{4}\frac{v^2}{c^2}\sin(2\theta) + \cdots$$

The aberration corrected vector is in the same plane as the source velocity vector and light time solution vector. The calculation of this vector from the aberration angle κ and the angle between the velocity vector and the vector from the observer to the source (θ) is illustrated on Fig. 3.5. The vector ρ_{in} is the light time solution from the observer to the source. The vector $\bar{\rho}$ is the direction that the source is observed and the direction that one would point a telescope. The angle κ between these vectors is the aberration angle as defined above. From the geometry, the vector

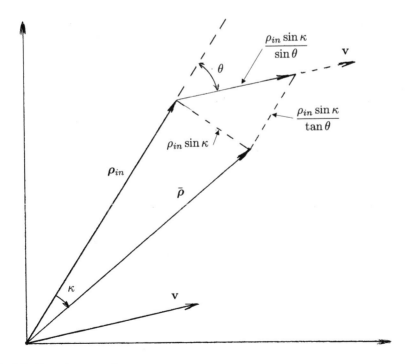

Fig. 3.5 Aberration correction

formula for planetary aberration is simply

$$\bar{\rho} = \left\{ \rho_{in} + \left[\frac{\rho_{in} \sin \kappa}{v \sin \theta} \right] \mathbf{v} - \left[\frac{\sin \kappa}{\tan \theta} \right] \rho_{in} \right\}$$

If the source is a star or remote object, the magnitude of the light time solution vector approaches infinity. The formula for stellar aberration ($\bar{\rho}_s$) may be obtained by taking the limit and

$$\bar{\rho}_s = \left\{ \hat{\rho}_{in} + \left[\frac{\sin \kappa}{v \sin \theta} \right] \mathbf{v} - \left[\frac{\sin \kappa}{\tan \theta} \right] \hat{\rho}_{in} \right\}$$

Chapter 4
General Relativity Time Delay Experiment

In February of 1997, the Near Earth Asteroid Rendezvous (NEAR) spacecraft passed behind the sun and continued on to encounter the asteroid Mathilde and rendezvous with the asteroid Eros. As the spacecraft passed behind the sun, the path of the two-way radio signal from the Deep Space Network (DSN) comes close to the sun and vanishes as the spacecraft is eclipsed by the sun. The signal from the DSN to the spacecraft and back to the Earth is delayed from what it would be if the sun was not present due to solar plasma, and the effect of general relativity. Interplanetary spacecraft are often eclipsed by the sun because planets, comets, and asteroids tend to have orbits near the ecliptic plane. However, the NEAR geometry was particularly favorable and provided an excellent opportunity to measure the general relativity delay. Since the data was readily available, it presented the opportunity to repeat the experiment performed on the Viking spacecraft in 1976. The only modification to navigation software was the addition of the capability to estimate the constant parameter γ of general relativity. If γ is zero, there is no general relativity, but if γ is one, Einstein is correct. The experiment during the Viking mission used range as the primary data type. On NEAR an experiment was proposed that would use Doppler as the primary data type. At the time, the Viking experiment was the most accurate verification of general relativity. The parameter γ was verified to an accuracy of about 0.1%. A justification for performing the NEAR experiment was to verify the operational navigation software to be used during the orbit phase of the mission.

The spacecraft trajectory shown schematically in Fig. 4.1 is far from the sun where the spacecraft is essentially in flat space. The Earth is also approximately in flat space. If the sun were not present, the triangle ADF would be nearly a right triangle that obeys Pythagoras's theorem. With the sun in place, the measured line segment from A to E, which passes the sun at 18 solar radii, would be about 14 km longer due to the curvature of space. The point D would be about 49 km longer if the spacecraft was visible. This violates the Pythagorean theorem, which would predict the square root of AF squared minus DF squared. The spacecraft trajectory

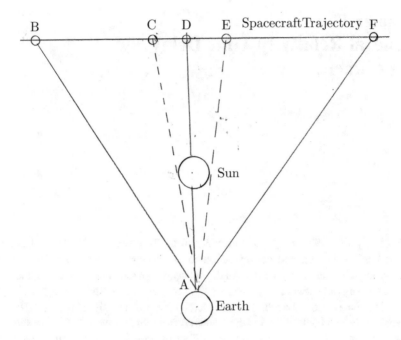

Fig. 4.1 General relativity experiment geometry

is approximately in a circle centered at the sun whose circumference is $2 Pi\ r$, where Pi is computed from the General Theory of Relativity.

4.1 Observed Path Length

The observed path length is the distance between the spacecraft and an antenna on the Earth. For convenience a planetocentric coordinate system is defined at the Earth's center of mass. The Earth antenna location is defined as the point on the antenna at the focus of the parabolic antenna dish. This point is the vector sum of the station location and an antenna correction that is computed from the direction that the antenna is pointed in a topocentric coordinate system. The station location is corrected for polar motion, solid Earth tides, and continental drift. The point on the antenna is corrected for cabling from the transmitter and the receiver. The transmitted and received signals pass through the troposphere and ionosphere and are delayed by a few meters. The transmitted and received signals are also delayed by charged particles near the sun. At the spacecraft, the received signal is turned around in a phase coherent loop so that the Doppler and range cycle count is not lost.

4.2   Computation of Observed Path Length

The metric tensor of general relativity is obtained by solution of the Einstein field equations. The solution for a massless particle in the gravity field of a single body was obtained by Schwarzschild and is given below in isotropic coordinates. The line element from the Schwarzschild metric tensor is given by

$$ds^2 = -\left(1 + \frac{2\gamma\mu}{c^2 r}\right)(dx^2 + dy^2 + dz^2) + \left(1 - \frac{2\gamma\mu}{c^2 r} + \frac{2\gamma\mu^2}{c^4 r^2}\right) c^2 dt^2$$

As shown in Eq. (3.15), the path length, neglecting terms greater than 1 ns, is given by

$$t_2 - t_1 = \frac{1}{c}\int_{z_1}^{z_2}\left[1 + \frac{\gamma\mu}{c^2\sqrt{R^2 + z^2}}\right] dz$$

$$t_2 - t_1 = \frac{z_2 - z_1}{c} + \frac{\gamma\mu}{c^3}\ln\left[\frac{z_2 + \sqrt{z_2^2 + R^2}}{z_1 + \sqrt{z_1^2 + R^2}}\right] \tag{4.1}$$

Here, γ is a constant that is zero if there is no general relativity and one if there is general relativity.

4.3   Troposphere Calibration

A radio signal passing through the Earth's troposphere will be delayed depending on the dielectric constant of the media and path length.

$$\Delta t^t = G_t(t, x, y)$$

The troposphere delay has been conveniently separated into wet and dry components that are functions of delay at zenith (z) and elevation angle (γ).

$$G_t(t, x, y) = R_d + R_w$$

The first term in the above equation represents the non-linearity of the dry troposphere mapping function, and the second term represents the variation in the dry troposphere z height due to local weather. The next two terms are the same quantities for the wet troposphere. The wet and dry troposphere mapping functions are tabulated as delay as a function of spacecraft elevation angle. Empirical formulas for these mapping functions are given by

$$R_d = \frac{z_d}{\sin \gamma + \dfrac{A_d}{B_d + \tan \gamma}}$$

$$R_w = \frac{z_w}{\sin \gamma + \dfrac{A_w}{B_w + \tan \gamma}}$$

where

$$\sin \gamma = \cos \delta \cos \lambda \cos \phi + \sin \lambda \cos \phi + \sin \phi \sin \delta$$

$$\lambda = \omega_e t + \lambda_s - \alpha$$

The dry component of the troposphere (R_d) is a function of the delay at zenith (z_d), the elevation angle (γ), and constants A_d and B_d that are provided to model the bending at low elevation angles. The wet component (R_w) is similarly defined. The elevation angle (γ) is computed as a function of the latitude of the tracking station (ϕ), the declination of the spacecraft (δ), and the local hour angle with respect to the spacecraft (λ). The local hour angle is zero when the spacecraft is at zenith and is a function of the Greenwich hour angle ($\omega_e t$), the station longitude (λ_s), and the right ascension of the spacecraft (α).

The troposphere dry component is assumed to be stable. Most of the variability is associated with the wet component. The variation in the wet component can be modeled as a periodic variation in the z height (z_w). The hourly variation in the wet component of the troposphere appears as a random walk that would require a high-order Fourier series to represent analytically. The variation may be modeled as a simple sinusoid with amplitude and frequency selected to be representative of the short-term variation.

$$z_w = z_{w_0} + z_{w_1} \sin(\omega_{w_1} t)$$

4.4   Plane Wave Propagation Through Ionized Gas

The following analysis demonstrates that the phase velocity of the Doppler signal exceeds the speed of light by the same amount the range signal is delayed. We first consider the phase velocity associated with Doppler data.

The propagation of a plane wave through an ionized gas such as the Sun's corona is described by Maxwell's equations, specifically the laws of Faraday and Ampere in the vector form:

$$\nabla \times \mathbf{E} = -\mu_0 \frac{\partial \mathbf{H}}{\partial t} \tag{4.2}$$

$$\nabla \times \mathbf{H} = \epsilon_0 \frac{\partial \mathbf{E}}{\partial t} + \mathbf{J} \tag{4.3}$$

If we assume a plane transverse wave in the z direction with associated electrical field $\mathbf{E} = E(z,t)\mathbf{i}$, magnetic field $\mathbf{H} = H(z,t)\mathbf{j}$, and current density $\mathbf{J} = J(z,t)\mathbf{i}$, Eqs. (4.2)–(4.3) reduce to the one-dimensional wave equation in E. The constants μ_0 and ϵ_0 are, respectively, the permeability and permittivity of free space.

$$\frac{\partial^2 E}{\partial z^2} = \mu_0 \epsilon_0 \frac{\partial^2 E}{\partial t^2} + \mu_0 \frac{\partial J}{\partial t} \tag{4.4}$$

Equations describing the motion of free electrons in a time varying electric field are also needed. The force (**F**) on an electron in an electric field (**E**) is proportional to the charge (*e*) and equal to the mass of the electron times its acceleration in the direction of (**F**):

$$\mathbf{F} = e\mathbf{E} = m \frac{d^2 \mathbf{r}}{dt^2} \tag{4.5}$$

The current density is the electron charge (*e*) times the electron flux or the number of electrons that pass through a given area per unit time. The electron flux is the product of the electron density (*N*) and the velocity of the electrons.

$$\mathbf{J} = Ne \frac{d\mathbf{r}}{dt} \tag{4.6}$$

For a plane transverse wave, these equations for the electron reduce to

$$m \frac{d^2 r}{dt^2} = eE$$

$$J = Ne \frac{dr}{dt}$$

which, when substituted into Eq. (4.3), result in the following wave equation:

$$\frac{\partial^2 E}{\partial z^2} = \mu_0 \epsilon_0 \left(\frac{\partial^2 E}{\partial t^2} + \frac{Ne^2}{m\epsilon_0} E \right) \tag{4.7}$$

A solution to this one-dimensional wave equation for an ionized atmosphere is

$$E = E_0 \sin [\omega t - kz] \tag{4.8}$$

from which we obtain

$$r = \left\{ \frac{-e}{\omega^2 m} \right\} E$$

$$J = \left\{ \frac{-Ne^2}{\omega^2 m} \right\} \frac{dE}{dt}$$

where the frequency is $\omega/2\pi$ and k is the so-called wave number that is obtained by substituting the solution into the one-dimensional wave equation given above. If we define $\omega_p^2 = \dfrac{Ne^2}{m\epsilon_0}$, we have

$$k^2 = \omega^2 \mu_0 \epsilon_0 \left\{ 1 - \frac{Ne^2}{\omega^2 m \epsilon_0} \right\} = \omega^2 \mu_0 \epsilon_0 \left\{ 1 - \frac{\omega_p^2}{\omega^2} \right\} \tag{4.9}$$

The phase velocity of the wave is defined by the locus of points along z where E_z is constant. Thus we have,

$$\omega t - kz = \text{constant}$$

which implies

$$v = \frac{dz}{dt} = \frac{\omega}{k}$$

When $N = 0$, we have the velocity of an electromagnetic wave in a vacuum and

$$v = c = \frac{1}{\sqrt{\mu_0 \epsilon_0}}$$

The wave length (λ) is related to the wave number (k), frequency (f), and phase velocity(v) by the following equation:

$$\lambda = \frac{v}{f} = \frac{2\pi v}{\omega} = \frac{2\pi}{k}$$

The phase velocity of the wave is thus

$$v = \frac{c}{\left(1 - \dfrac{\omega_p^2}{\omega^2} \right)^{\frac{1}{2}}} \tag{4.10}$$

The phase velocity in a dispersive medium is always greater than the speed of light. This apparent contradiction of special relativity is possible because the radio signal phase velocity does not describe the actual velocity of mass or energy, but rather the velocity of a pattern, or mathematical entity.

Since it is critical to our analysis that ω_p/ω be less than 1.0, it is helpful to estimate it at this time. For this it is necessary to estimate N, the electron density in the plasma, which depends on the distance of the signal path from the Sun. Because the NEAR spacecraft is eventually occulted by the Sun, the closest approach distances of the signal path go to 0, but at less than 18 solar radii, the signal was degraded beyond usability. At 18 solar radii, previous estimates have placed N at on the order of 10^3 electrons per cubic centimeter, which yields an ω_p of less than 1.0 MH. For an X-band signal, $\omega = 2\pi\ f$, where f is approximately 8.6 GH. Thus (ω_p^2/ω^2) is small, on the order of 10^{-8}. Since N decreases with increasing distance from the Sun, this is an upper bound. The data analyzed in this chapter includes signal paths with the closest approach distances of from 40 to 18 solar radii, taken over approximately 1 week.

We next consider the delay in the group velocity, which is associated with the range data. The concept of group velocity arises when we have electromagnetic waves that are nearly the same frequency traveling in the same direction through the same medium. Because linearity holds for electromagnetic waves, any electromagnetic wave may be regarded as the sum of its individual frequency components. Consider the case of two electromagnetic waves that differ in frequency and wave number by an infinitesimal amount, $\delta\omega$ and δk, respectively. When added together, we obtain the wave packets illustrated below (Fig. 4.2).

The resultant wave is the higher frequency carrier that moves with phase velocity v as described above and the wave packets formed by the beating of the two nearly equal in frequency waves that move at a different group velocity (u). Doppler tracking data is associated with the phase velocity and range data is associated with the group velocity. In a vacuum, the phase and group velocities are equal to the

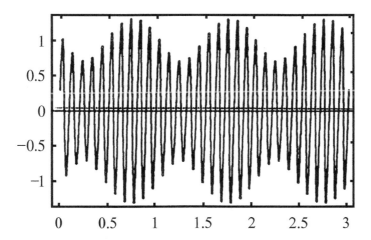

Fig. 4.2 Wave packets

speed of light. The two electromagnetic waves alluded to above are given by

$$\psi_1 = \sin(\omega t - kz)$$

$$\psi_2 = \sin[(\omega + \delta\omega)t - (k + \delta k)z]$$

We must perturb both the frequency and wave number in order to get the correct velocity that is controlled by the medium. The resultant wave is obtained by adding the two electromagnetic waves. After some trigonometric substitutions, we have

$$\psi = \psi_1 + \psi_2 = 2\cos\left(\frac{\delta\omega t - \delta k z}{2}\right)\sin\left[\left(\omega + \frac{\delta\omega}{2}\right)t - \left(k + \frac{\delta k}{2}\right)z\right] \quad (4.11)$$

The carrier is given by the sine term and the modulation of the carrier is given by the cosine term. In a dispersive medium, the carrier wave moves at a velocity greater than the speed of light as shown above, and the wave packet described by the cosine term moves at a velocity slower than the speed of light. The velocity of the wave packet is obtained in the same manner as described above for the carrier. The locus of points along z where the amplitude of the wave packet is constant is given by

$$\frac{\delta\omega t - \delta k z}{2} = \text{constant}$$

The group velocity is thus

$$u = \frac{dz}{dt} = \frac{\delta\omega}{\delta k} = \frac{d\omega}{dk}$$

Differentiating Eq. (4.12) for the wave number, we get the desired group velocity.

$$u = \frac{k}{\omega}c^2 = c\left(1 - \frac{\omega_p^2}{\omega^2}\right)^{\frac{1}{2}}$$

Because the group and phase velocities are close to the speed of light, ω_p is small, and we may make the following approximations:

$$v \approx c\left(1 + \frac{\omega_p^2}{2\omega^2}\right) \qquad (4.12)$$

$$u \approx c\left(1 - \frac{\omega_p^2}{2\omega^2}\right) \qquad (4.13)$$

The first is accomplished by expanding v in a geometric series, completing the square of the first two terms, and neglecting terms of fourth order and higher. The second approximation is obtained by completing the square and neglecting higher order terms. We obtain the result that the delay in the range measurement associated with charged particles is approximately equal to the advance in the Doppler measurement and that these two measurements, when obtained simultaneously, may be used to calibrate the effect of charged particles on the total path length.

4.5 Ionosphere Calibration

A radio signal passing through the ionosphere experiences a reduction in group velocity and an equal increase in phase velocity that is a function of the frequency and the number of charged particles along the signal path. The Doppler measurement is dependent on the phase velocity, and the advance of the signal is functionally defined by

$$\Delta t^i = G_i(t, x, y)$$

An empirical formula for the effect of the ionosphere on the Doppler measurement is given by

$$G_i = \frac{-1}{c} \sum_{j=0}^{n} k\, C_j\, X^j$$

$$X = 2\left(\frac{t - t_a}{t_b - t_a}\right) - 1$$

where the C_js are coefficients of a polynomial in time (t) from t_a to t_b normalized over the interval of -1 to $+1$ and k is a proportionality factor.

4.6 Solar Plasma Time Delay

The range delay associated with a plane wave passing through the Sun's corona is obtained by integrating the group velocity of propagation along the path length.

$$\int_{t_1}^{t_2} dt = \int_{z_1}^{z_2} \frac{dz}{u_z} = \int_{z_1}^{z_2} \frac{1}{c}\left[1 + \frac{e^2}{2m\epsilon_0\omega^2} N(z)\right] dz \tag{4.14}$$

The electron density varies approximately as the inverse square of the distance from the Sun.

$$N(z) = \frac{N_0 r_s^2}{r^2} = \frac{N_0 r_s^2}{R^2 + z^2} \tag{4.15}$$

R is the perpendicular distance from the center of the Sun to the light path or the distance of closest approach to the Sun. The constant N_0 is the effective electron density at the surface of the Sun and r_s is the radius of the Sun. Carrying out the integration, we have for the delay

$$t_2 - t_1 = \frac{z_2 - z_1}{c} + \frac{e^2}{2cm\epsilon_0\omega^2} \frac{N_0 r_s^2}{R} \left\{ \tan^{-1}\left(\frac{-z_1}{R}\right) + \tan^{-1}\left(\frac{z_2}{R}\right) \right\} \qquad (4.16)$$

The time advance of the Doppler signal, which is associated with the phase velocity, is the same equation as above except with a minus sign.

4.7   Doppler Measurement

Doppler data provides a direct measure of line-of-sight velocity of a spacecraft relative to a tracking antenna. The accuracy of this measurement is about 1 mm/s when the two-way Doppler count is integrated for 1 min. A single Doppler measurement provides no information on position or velocity normal to the line-of-sight. For those phases of the mission where the spacecraft is being accelerated rapidly, such as near a planetary encounter, a series of Doppler measurements permit a very accurate complete orbit determination by observing the orbit dynamics signature. When the spacecraft is far from a planet, comet, or asteroid, the gravitational accelerations are not sufficient to observe this signature. However, the "velocity parallax" due to the tracking stations rotation with the Earth provides a measure of position normal to the line-of-sight. By measuring the amplitude and phase of the tracking stations signature, the right ascension and cosine of declination can be determined to about $0.25\,\mu$rad. Thus, at Jupiter distance, the Earth-relative orbit determination error is about 200 km.

The functional definition of Doppler data as line-of-sight velocity is useful for analyzing the orbit determination errors that are spacecraft or trajectory dependent but is of little use for analyzing error sources close to the actual measurement such as media or hardware errors. The actual measurement is a count derived from the signal received from the spacecraft and a frequency standard maintained at the tracking station that controls the frequency of the transmitted signal. Thus, a precision model of the Doppler observable includes a model of the signal path as well as hardware elements. In practice, the hardware errors are small compared to media, station location, and spacecraft dynamics errors.

A model of the Doppler observable has been developed that idealizes some of the hardware error sources yet precisely models the external environment. This model is sufficiently precise for computation of the observable and is essentially the model contained in orbit determination software. Of particular interest are models that are external to the tracking station hardware yet pertain directly to the signal path. Media and the effect of general relativity on station clocks are examples. Other models, such as station locations and polar motion, though not directly part of the Doppler measurement system, may be treated as measurement calibrations.

The Doppler measurement is simply an electronic count of the number of cycles from a frequency standard (N_c) minus the number of cycles of the spacecraft signal received by the tracking station (N_r) and scaled by the count time interval (ΔT_c). Thus we have

$$Z_m = \frac{(N_c - N_r) + n}{\Delta T_c}$$

where n is the measurement noise that is about 1/10 of a cycle. The received frequency and standard frequency need not be counted individually and differenced but may be added together electronically and the beat frequency counted. This is a detail that is dependent on the hardware implementation. The numerical value of Z_m is the number that is recorded on a tracking data file and used for orbit determination.

In the orbit determination software, we need to obtain a computed value for Z_m as a function of parameters that are available. This function can be derived from the equations of motion and a physical model of the system. We start by developing a frequency standard that can be compared with the frequency of the transmitted and received signals. The frequency standard is obtained by scaling a reference oscillator frequency f_q, obtained from an atomic clock, to equal the transmitted frequency (f_t) times the spacecraft turn around ratio (C_3), which would nominally be the received frequency if there were no spacecraft Doppler shift or additional delay. The turn around ratio is necessary so that the downlink will not interfere with the uplink.

$$N_c = C_3 \, f_t \, \Delta T_c$$

where for S-band Doppler,

$$C_3 = \frac{240}{221}$$

$$f_t = 96 \, f_q$$

$$\Delta T_c = T_{3_e} - T_{3_s}$$

The count time (ΔT_c) is defined as the difference between the reception time at the start of the count time interval (T_{3_s}) and the reception time at the end of the interval (T_{3_e}). For a schematic representation of these times, see Fig. 4.3. In the above equation, all of the parameters are constant or arbitrarily specified including the reception times. The real information content of the measurement is contained within the count N_r. Thus, in order to obtain a complete equation for the computed measurement, we need an equation for N_r. It is tempting to differentiate and work in the frequency domain; however, the hardware works with phase which makes

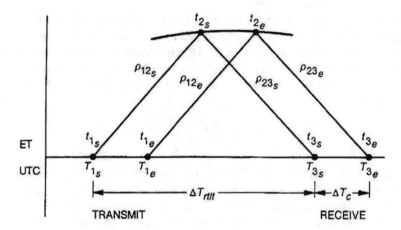

Fig. 4.3 Doppler observable schematic diagram

it convenient to formulate the measurement in terms of phase, thus bypassing an explicit equation for the received frequency.

The equation that relates the measurement to the observable parameters is

$$N_r = \frac{240}{221} N_t$$

where

$$N_t = f_t \, (T_{1_e} - T_{1_s})$$

$$f_t = 96 \, f_q$$

The above equation for N_r states that the number of cycles counted at the receiver is equal to the number of cycles transmitted N_t times the spacecraft turn around ratio. This equation is true because they are effectively the same cycles. Thus, the information content of the measurement is now contained in the transmit times T_{1_e} and T_{1_s}. Since both of these times are unknown, we need some additional equations to tie into the observable quantities. At this point in the development, we have the following equation for the computed measurement:

$$Z_c = (T_{3_e} - T_{3_s} - T_{1_e} + T_{1_s}) \, \frac{C_3 f_t}{\Delta T_c}$$

We need equations for the times in the above equation. They will be developed as functions of ephemeris time t. We have for the atomic clock at the station

$$T = t + F(t, x, y)$$

The station time T is equal to the ephemeris time t modified by a small correction due to general relativity and any other parameter that may affect the running of the clock. The calibration function (F) is a function of t, the state of the solar system (x), and other constant parameters (y). Here, x and y can be thought of as parameter vectors. The relevant times shown in Fig. 4.3 relating to the Doppler measurement are

$$T_{1_s} = t_{1_s} + F(t_{1_s}, x, y)$$

$$T_{1_e} = t_{1_e} + F(t_{1_e}, x, y)$$

$$T_{3_s} = t_{3_s} + F(t_{3_s}, x, y)$$

$$T_{3_e} = t_{3_e} + F(t_{3_e}, x, y)$$

Making the above substitutions, the equation for the computed measurement becomes

$$Z_c = (t_{3_e} - t_{3_s} - t_{1_e} + t_{1_s}) \frac{C_3 f_t}{\Delta T_c}$$

$$+ \left[F(t_{3_e}, x, y) - F(t_{3_s}, x, y) - F(t_{1_e}, x, y) + F(t_{1_s}, x, y) \right] \frac{C_3 f_t}{\Delta T_c}$$

$$(4.17)$$

Since the speed of light is constant in any reference frame, by integrating along the light path, we obtain

$$t_{3_e} - t_{1_e} = \frac{\rho_{12_e} + \rho_{23_e}}{c} + \Delta t_{1_e}^m + \Delta t_{3_e}^m$$

$$t_{3_s} - t_{1_s} = \frac{\rho_{12_s} + \rho_{23_s}}{c} + \Delta t_{1_s}^m + \Delta t_{3_s}^m$$

where the ρ terms represent the integrated distance along the light path and the t^m terms represent the additional delay caused by media. The distances along the light path are obtained by integrating the equations of motion.

$$\rho_{12_s} = \int \int_{t_{1_s}}^{t_{2_s}} \ddot{\rho} \, dt dt$$

$$\rho_{23_s} = \int \int_{t_{2_s}}^{t_{3_s}} \ddot{\rho} \, dt dt$$

$$\rho_{12_e} = \int \int_{t_{1_e}}^{t_{2_e}} \ddot{\rho} \, dt dt$$

$$\rho_{12_e} = \int \int_{t_{2_e}}^{t_{3_e}} \ddot{\rho} \, dt dt$$

These equations are referred to as the light time equations and are solved iteratively for the arguments of integration. The media delay is included in the measurement equation by evaluating the calibration function (G) at the appropriate times.

$$\Delta t^m = G(t, x, y)$$

$$\Delta t^m_{1_s} = G(t_{1_s}, x, y)$$

$$\Delta t^m_{1_e} = G(t_{1_e}, x, y)$$

$$\Delta t^m_{3_s} = G(t_{3_s}, x, y)$$

$$\Delta t^m_{3_e} = G(t_{3_e}, x, y)$$

The final equation for the computed measurement includes the observable equations as well as clock and media calibration functions.

$$Z_c = \frac{\rho_{12_e} + \rho_{23_e} - \rho_{12_s} - \rho_{23_s}}{c} \frac{C_3 f_t}{\Delta T_c}$$

$$+ \left[F(t_{3_e}, x, y) - F(t_{3_s}, x, y) - F(t_{1_e}, x, y) + F(t_{1_s}, x, y) \right] \frac{C_3 f_t}{\Delta T_c}$$

$$+ \left[G(t_{3_e}, x, y) - G(t_{3_s}, x, y) + G(t_{1_e}, x, y) - G(t_{1_s}, x, y) \right] \frac{C_3 f_t}{\Delta T_c}$$

$$(4.18)$$

4.8 Range Measurement

The range data type provides a measure of the range between a DSN station and a spacecraft. The range is inferred from the time it takes a radio signal to travel from the DSN station to the spacecraft and back to the station. The radio signal is transmitted to the spacecraft where it is received and retransmitted back to the tracking station. The round trip light time is determined by impressing a pattern, referred to as a range code, on the transmitted carrier and detecting this pattern in the received radio signal. The range code provides time markers in the transmitted and received radio signal that may be measured with high precision by an atomic clock. For orbit determination, a computed value of the measurement is obtained from a mathematical model similar to the model used for Doppler data.

Range data has essentially the same information content as Doppler data. Range data provides the integral of Doppler data over some time interval. This integrated Doppler can be determined by differencing two range measurements. The integrated Doppler is more accurate than differenced range. However, the range data provides the constant of integration. Doppler data alone determines range through the orbit dynamics. For this reason, an orbit determination strategy has evolved to process a single loosely weighted range point for each station pass to initialize the Doppler. Processing range and Doppler together at the same weight can result in aliasing. Both data types determine the right ascension and declination independently and they may disagree.

The first step in computing the two-way range observable is to assemble all the input data required by the orbit determination software. These are essentially the same models as used for Doppler data with some minor exceptions. Models of the transmission media, station locations and the effect of general relativity are virtually the same as used for computing the Doppler observable. One exception is the sign of delays associated with charged particles in the ionosphere and solar plasma. For Doppler data, the charged particles speed up the velocity of the carrier and the delay is subtracted. For range data, the velocity of the carrier is slowed down, and the delay is added as for all the other media delays. The magnitude of the velocity increase associated with charged particles for Doppler data is equal to the velocity decrease or delay for range data.

A range data point is read from the tracking data file to obtain the time tag ($TIMTAG$), frequency ($FRQCY$), the lowest ranging component ($NLOW$), and the measurement ($ROBS$). The light time equation is solved for the transmit and spacecraft times t_1 and t_2. The station receive time (t_3) is equal to $TIMTAG$.

The next step is to integrate the ramp tables for the uplink and downlink. The ramp tables keep the received signals in the center of the carrier bandwidth. Since the range code is modulated on the carrier, the range traveled by the radio signal equals the sum of the wavelengths associated with all the cycles between the spacecraft and the DSN antenna and is equal to the cycle count times the speed of light after correcting for media. The cycle count (f_q) is obtained by integrating the uplink ramp table from t_1 to t_3. The range observable (Z_r) is computed from the output of the ramp table integration which is scaled by an integer ratio corresponding to the frequency dividers used in the actual hardware implementation to give the measurement in range units (R_u). For S-band frequency, the conversion to range units is

$$R_u = \frac{1}{2} fq$$

For X-band frequency, the conversion for the 34-m Az-EL high-efficiency antenna (HEF) is

$$R_u = \frac{11}{75} fq$$

and for 34-m Block 5 Receivers (BVR)

$$R_u = \frac{221}{749 \times 2} fq$$

The range code is a pattern consisting of square waves whose frequency decreases by powers of two. Thus, the range code is repeated at a rate determined by the lowest frequency square wave or range component. This results in an ambiguity in the determination of range that must be resolved by introducing information from other sources, most notably the Doppler measurement that has no ambiguity. The range ambiguity manifests itself as a roll over to zero in the range unit counter. Thus, if the computed number of range units is greater than the ambiguity, the ambiguity is repeatedly subtracted from the computed measurement until it is in the proper range. The range ambiguity is computed from $NLOW$, which is obtained from the tracking data file and is given by

$$AMBIG = 2^{(NLOW+6)}$$

The number of roll overs of the range unit counter is the integer part of

$$N_a = Integer \left(\frac{R_u}{AMBIG} \right)$$

The adjusted value for the range unit count is then

$$R_{ua} = R_u - N_a\, AMBIG$$

At the time the computed observable rolls over, it cannot be determined from the computed range alone whether the actual observable has just rolled over or is about to roll over. This ambiguity may be resolved by inspecting the measurement residual. The measurement residual is simply

$$RESID = ROBS - R_{ua}$$

where $ROBS$ is obtained from the tracking data file. If the absolute value of $ROBS$ is greater than $1.5 \times AMBIG$, the ambiguity resolution is skipped. Otherwise, the following adjustment is made to the computed observable:

$$If(RESID > 0.5 \times AMBIG)\ Z_r = R_{ua} + AMBIG$$

$$If(RESID < 0.5 \times AMBIG)\ Z_r = R_{ua} - AMBIG$$

Otherwise, $Z_r = R_{ua}$. The residual (RESID) is then recomputed with the new value for Z_r.

$$RESID = ROBS - Z_r$$

This correction to the range measurement can be dangerous. If the ambiguity is set too low and the computed measurement is not known within the ambiguity, the range measurement will be in error and the resultant orbit solution can be off by several hundred kilometers. The range and Doppler residuals could be flat making it difficult to detect this error.

4.9   Experimental Results

X-band Doppler and range measurements from 3 Deep Space Network stations from October 1, 1996 to March 7, 1997 were obtained from the NEAR mission. The X-band signal frequencies were 8.9 GHz uplink and 7.8 GHz downlink. Although Doppler data was processed from all the three stations, station 15 (Goldstone, California), station 45 (Canberra, Australia), and station 65 (Madrid, Spain), in order to ensure a precise trajectory estimate, only the range data from station 15 was used in order to eliminate station location bias. One-sigma errors in the Doppler and range are known to be within 0.1 mm and 0.7 m, respectively.

The range data became inaccessible at signal path distances from the Sun closer than 18 solar radii. The ranging signal consists of a sequence of square waves of varying frequencies, from 1.0 MHz to 1.0 Hz in the case of the NEAR mission. The highest frequency is the clock. Additional frequencies decrease by powers of two. In a dispersive medium, higher frequencies are shifted more than lower frequencies, causing a blurring of the square wave, until the range signal can no longer be resolved. In hindsight, it would have been beneficial to send range signals at a lower clock frequency as the signal path passed close to the Sun, in order that the range signal be available at closer distances.

The data was processed by a Square Root Information Filter (SRIF) to estimate the parameters that describe the spacecraft trajectory during the solar conjunction period. The estimated parameters were spacecraft state, a single propulsive maneuver, solar radiation pressure, three components of stochastic acceleration, solar plasma electron density, and γ of general relativity. The electron density parameter (N), the coefficient for the inverse square term of the solar plasma model, and γ were estimated as stochastic parameters. Even though these parameters are constant, they were allowed to vary as a function of time by the filter. For each data batch, solutions were obtained for electron density and γ. It is possible to separate these parameters by using the group delay associated with range to calibrate the phase advance associated with Doppler caused by the solar plasma, while the general relativity delay is the same for both Doppler and range.

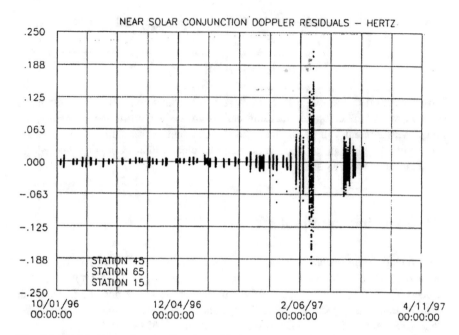

Fig. 4.4 NEAR solar conjunction Doppler residuals

Figure 4.4 shows the Doppler residuals associated with fitting data for the NEAR solar conjunction of February 1997. The data arc started on October 1, 1996 and extended through March 7, 1997. This long arc of data was necessary to enable solution for the spacecraft orbit as the spacecraft passes behind the Sun at a distance of about 3 AU from the Earth. The spacecraft, Sun, and Earth are close to being in alignment during this time interval. Since the range and Doppler measurement is close to this same direction, the problem of determining the delay associated with the radio signal as the ray path approaches the Sun is effectively reduced to one dimension. The other dimensions enter into the determination of the total path length in the second order. The usable data was restricted to the times when the NEAR spacecraft was held in an attitude that pointed the solar panels directly at the Sun. This attitude resulted in a steady solar pressure acceleration that could be modeled with high precision. Early in March 1997, normal spacecraft attitude resumed and the data was no longer usable. During the approach, there was one spacecraft propulsive maneuver that also tended to corrupt the data arc. As the ray path approached the Sun, the Doppler noise increased to the point where the data became marginally useful. This data was deleted from the solution. The absence of range data when the ray was within about 18 solar radii also rendered the Doppler data as marginally useful.

Figure 4.5 shows pre-fit range residuals for the period of time of interest for orbit determination during the NEAR solar conjunction of February 1997. All of the data that was acquired is shown here. Each grouping of points represents a station pass.

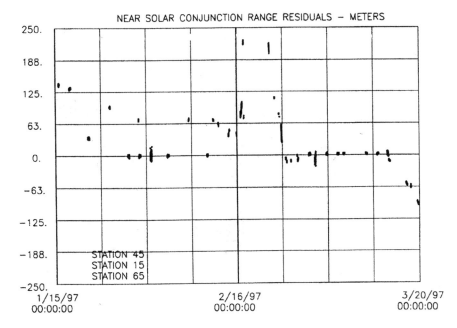

Fig. 4.5   NEAR solar conjunction pre-fit range residuals

Because of range calibration errors caused by the noisy environment associated with solar plasma, many of the station passes were excluded from the final solution. The bad range passes are caused by dispersion associated with the solar plasma, which, when present in the data, produces jumps in the range measurement. The good range passes have noise at the 20 m level, and the variation in range bias is only a few meters. This characteristic of the data makes it possible to identify the bad passes and eliminate them.

Figure 4.6 shows the range residuals after the data has been fit and the bad data removed. The noise is only about 3 m for nine passes, and for one pass, the noise is about 10 m. Averaging the noise over the encounter time interval, the range measurement error should be below 1.0 m. Since the total delay associated with general relativity is on the order of 20 km, it may be possible to design an experiment to measure the parameter γ of general relativity to an accuracy of less than 0.01%. Achieving this level of accuracy would require reducing all the model errors to the same accuracy as the range measurement error. With current orbit determination techniques, the error in determining γ is about an order of magnitude above the theoretical limit.

Figure 4.7 shows the one-way delay in meters for various corrections that are applied to the computed range and Doppler measurements. The delay in meters is the time delay in seconds multiplied by the speed of light. Meters are preferred for quantifying these delays because they enter directly into the computation of the path length from the DSN antenna to the spacecraft antenna. The delays

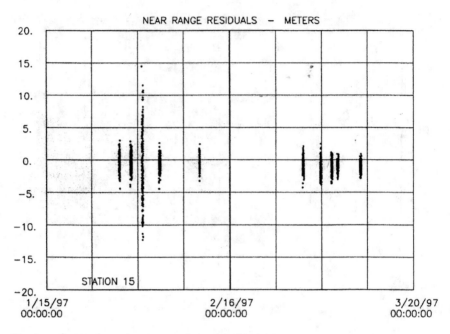

Fig. 4.6 NEAR solar conjunction post-fit range residuals

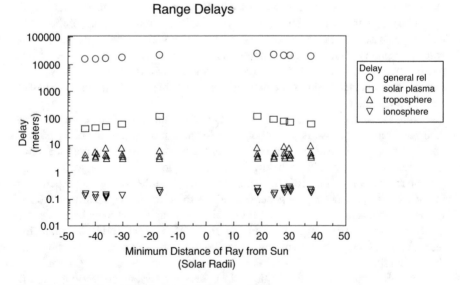

Fig. 4.7 NEAR solar conjunction range calibration delays

are those associated with the ionosphere, troposphere, solar plasma, and general relativity and are included in both the uplink and downlink signals. The total delay is approximately twice the one-way delay. They are not exactly the same, because of the motion of the antenna during the round trip signal time, which is about 50 min during the period of solar conjunction, and because of the different frequencies of the uplink and downlink carriers. Also, the effect of the charged particle environment is to actually advance the Doppler signal, so the distance associated with the delay is subtracted from the computed path length.

The range delays shown in Fig. 4.7 are given as a function of the minimum distance R from the Sun to the ray path between the DSN station and the spacecraft. Negative values are associated with the time period before conjunction during which the minimum distance is decreasing. The absolute range measurement from the DSN antenna to the spacecraft is only marginally useful because of range biases. The ionosphere delays average about 10 cm and the troposphere delays vary from about 3 to 10 m. They are determined by measurements taken at each DSN station. The variation in the troposphere delay is caused by the elevation angle above the horizon. Its effect may be minimized by editing low elevation angle data out of the solution. Both the solar plasma and general relativity delays increase as the minimum distance of the ray path to the Sun decreases. The signal completely disappears at one solar radius. The two-way general relativity delay is about 250 μs at the point where the ray grazes the Sun, which corresponds to about 36 km in the one-way delay. At 18 solar radii, the minimum distance for which we have usable range data from the NEAR mission, the one-way delay is about 14 km (14,000 m on Fig. 4.7) and decreases to about 10 km at 40 solar radii. It is this variation in the distance, associated with the curvature of space, that enables us to determine γ. It should be noted that the solar plasma delay increases dramatically as the ray path minimum decrease to a few solar radii, which makes this close in data less useful for spacecraft navigation, telemetry, and relativity experiments.

The plot shown in Fig. 4.8 shows the electron density profile as a function of distance from the sun for the solar eclipse of 1963, Mariner 6, Viking, and NEAR. The a priori value for noise in the stochastic model was 1000 electrons per cubic centimeter. This is 100% of the value observed during the solar eclipse of 1963. This value is consistent with the uncertainties in the values computed for Mariner 6 and Viking. The solar eclipse data was used as a priori because this determination is independent of any assumptions for γ of general relativity. The NEAR results are consistent with the previously determined values but are somewhat higher than those from the Viking mission. It is not known whether these differences are attributable to actual variations in the Sun's corona electron density or experimental error.

Figure 4.9 shows the variation in the estimated value of γ from general relativity theory as a function of time. We know from general relativity theory that γ is a constant parameter. By allowing the value of γ to vary as a function of time, we introduce conservatism into our estimation process and allow for the possibility that some other physical process is the source of the delay. If we get constant value for γ that is equal to one, we have a very strong confirmation of general relativity theory. The formal error we get from the SRIF estimates is about 0.5%, which

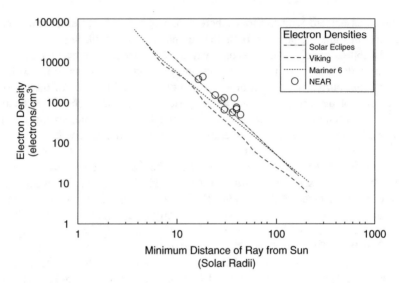

Fig. 4.8 Electron density profile comparison

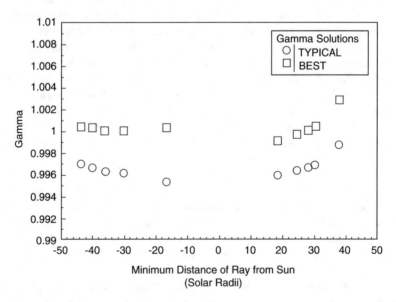

Fig. 4.9 Gamma as stochastic parameter

may be a bit pessimistic. Shown in Fig. 4.9 is the result of two filter runs. The one labeled "BEST" is from the run setup that is believed to most faithfully model the actual physical environment. Adjustments in the a priori assumptions will generate different solutions that are generally within 0.5% (γ between 0.995 and 1.005). Also shown in Fig. 4.9 is a typical solution that varied the a priori assumptions. Here, we read the Earth ephemeris from a file rather than integrating the Earth ephemeris from initial conditions. It is believed that the integrated Earth ephemeris should yield better results and that appears to be the case. However, this conclusion requires more study before we can definitively commit to either solution. The temptation is to pick the solution that most agrees with Einstein.

References

1. A. Einstein, The foundation of the general theory of relativity. Ann. Phys. **49**, 769–822 (1916)
2. A.S. Eddington, *The Mathematical Theory of Relativity* (Cambridge University Press, Cambridge, 1923)
3. H. Lass, *Vector and Tensor Analysis* (McGraw-Hill, New York, 1950)
4. I.S. Sokolnikoff, *Tensor Analysis Theory and Applications to Geometry and Mechanics of Continua* (Wiley, New York, 1964)

Index

Printed in the United States
by Baker & Taylor Publisher Services